Fashion

FROM CONCEPT TO CONSUMER

FOURTH EDITION

Gini Stephens Frings

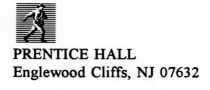

PRENTICE HALL
Englewood Cliffs, NJ 07632

Library of Congress Cataloging-in-Publication Data
Frings, Gini Stephens
    Fashion from concept to consumer / Gini Stephens Frings, — 4th ed
      p.   cm.
    Includes index.
    ISBN 0-13-177478-6
    1. Fashion. I. Title
TT518.F74 1994
687—dc20   93-43334
                             CIP

Editorial/production supervision: Eileen M. O'Sullivan
Acquisitions editor: Elizabeth Sugg
Page layout and interior design: Laura Ierardi
Cover design: Laura Ierardi
Production coordinator: Ed O'Dougherty
Page layout assistance: Stephen Hartner
Marketing manager: Debbie Sunderland
Editorial assistant: Maria Klimek
Production assistant: Bunnie Neuman

Cover sketch courtesy of Gianfranco Ferré

 © 1994, 1991, 1987, 1982 by Prentice-Hall, Inc.
A Simon & Schuster Company
Englewood Cliffs, New Jersey 07632

Prentice Hall would like to thank the following for their expert advice:

Guy Adamo	Jeffy Rosenhall
Berkeley College of Business	Philadelphia College of Textiles
Waldwick, NJ	Philadelphia, PA
Stella Warnick	Marcia Silberstein
Seattle Pacific University	FIDM
Seattle, WA	San Francisco, CA
Carolyn Blount	Nancy Cassill
Shoreline Community College	University of North Carolina–Greensboro
Seattle, WA	Greensboro, NC

Printed in the United States of America

10 9 8 7 6 5 4 3 2

ISBN 0-13-177478-6

Prentice-Hall International (UK) Limited, *London*
Prentice-Hall of Australia Pty. Limited, *Sydney*
Prentice-Hall Canada Inc., *Toronto*
Prentice-Hall Hispanoamericana, S.A., *Mexico*
Prentice-Hall of India Private Limited, *New Delhi*
Prentice-Hall of Japan, Inc., *Tokyo*
Simon & Schuster Asia Pte. Ltd., *Singapore*
Editora Prentice-Hall do Brasil, Ltda., *Rio de Janeiro*

For
Philipp,
Peter,
and
Victoria

CONTENTS

2

CONSUMER DEMAND   31

3

FASHION CHANGE AND CONSUMER ACCEPTANCE   53

4

FASHION RESEARCH AND ANALYSIS   67

Part Two

THE RAW MATERIALS OF FASHION 81

5

TEXTILE FIBER AND FABRIC PRODUCTION 83

Part Three

THE MANUFACTURING OF FASHION   139

8
INTERNATIONAL FASHION CENTERS   141

9
PRODUCT AND DESIGN DEVELOPMENT   163

10

APPAREL PRODUCTION   189

11

ACCESSORY AND FUR MANUFACTURING

217

12

THE GLOBAL MARKETPLACE

239

Part Four

THE RETAILING OF FASHION   265

13

RETAIL STORES   267

14

RETAIL FASHION MERCHANDISING   290

15

RETAIL FASHION PROMOTION   311

Appendix One

CAREER GUIDELINES   333

Appendix Two

THE BUSINESS OF FASHION   343

FOREWORD

Everyone has to wear clothes. Clothes are an important part of our lives. The clothing we buy for special occasions—proms, graduations, weddings, and so on—becomes part of our memories. We remember these occasions by what we were wearing.

This is what makes my business special. I enjoy designing those special dresses that my customers will remember. My customer knows what she wants. She wants to look beautiful. She wants to feel nostalgic about her purchase. I design with my customer in mind. My success comes from my strong focus, my point of view. I developed a special style and stuck to it over the years. I adapt current trends to my own style. Buyers know they can find a consistent look from me and so do my customers. I try every day to make my designs better than yesterday. I am never satisfied with myself. I have always felt that I have a special talent, a special creative force, but aside from that, I'm my own toughest critic.

Fashion design is not work to me, it is my fun, my life. I'm up at six, to work by eight, break half an hour for lunch—usually at my desk to look at fashion magazines so I can keep my eye on what's happening—then spend the rest of the day in the workroom making sure my designs are going the way I want. I create 150 designs a season, five times a year for our seven lines. But it's not enough to be creative. I have to combine my creative ability with a business sense, too. I love my business. No, it's more than that. What I do for a living is exactly the way I want to live.

I have known the author of this book for over 18 years. In her role as an educator, Gini would bring her students to meet with me to discuss my fashion philosophy and see my current line. Then the students would design and make a sample garment for the next season and bring them back to me for a critique. One of Gini's students later became my assistant.

Because of people like Gini, who has experience in the industry, fashion education is much better today. I am happy to introduce her excellent book which prepares the student to know how the fashion business operates and what to expect working in it. The book covers the fashion business in logical sequence with complete and realistic information. I hope each student will take advantage of this text and absorb its contents for later use.

As a student, you have to realize that your college education is just the beginning, a time to open your mind to new possibilities. Students want to be successful the minute they graduate and enter the field. But you can't expect to be a success overnight. As a graduate you must be focused and use your first job as a new learning situation and grow in your knowledge day by day.

Best wishes in your fashion career,

Jessica McClintock

Jessica McClintock
San Francisco

PREFACE

The purpose of this book is to tell the whole story of how the fashion business works, in sequential order from concept to consumer. The fashion business is an integrated series of processes from buying supplies, creating and developing a new product, and marketing the product. This process is repeated on three levels: raw materials, manufacturing, and retailing. The fashion business includes all the firms involved with producing apparel and accessories: material producers; garment and accessory manufacturers; and the retail stores that sell the finished product to the public. It is important for executives in the fashion industry to know how all the facets interrelate.

Fashion designers and manufacturers must work with textile producers to develop fabrics that they need for their garments. Fashion designers and manufacturers must also understand the importance of selling garments on the retail level. Retail fashion buyers should understand how garments are designed so that they can be creative merchandisers and make wise buying decisions.

Part One concentrates on fashion fundamentals. Chapter 1 traces the development of fashion and the fashion industry as a background to understanding today's business. Chapter 2 discusses influences on consumer demand that affect decisions at all levels of the industry. Chapter 3 explains fashion change and consumer acceptance. Chapter 4 covers market research, design analysis and design resources as a background for the designer and merchandiser.

Part Two covers the development, production, and marketing of raw materials, including textiles, trimmings, leather, and fur—the supplies needed for fashion manufacturing.

Part Three traces the fashion manufacturing process through design and merchandising development, production, and sales. The first test of a successful fashion design is at the wholesale market, the meeting ground of the manufacturer and retailer.

Part Four covers retailing: types of retail organizations, the buying and selling process, and promotion.

Each chapter contains list of objectives, review questions, terminology, and projects to aid in reviewing the subject matter. The appendices contain information on career guidelines and business organization. A glossary of fashion terminology is provided for easy reference.

This book will be a valuable tool for an introductory course in fashion design, marketing, merchandising, retailing, or business. There is also important information for textile marketing, apparel manufacturing, accessory design, production and marketing; and advertising and promotion. This is a text for specialists as well as those who are taking only a single course in fashion. In fact, it will interest anyone who wants to know more about fashion and the fashion business.

ACKNOWLEDGMENTS

I wish to thank the many friends and business associates who took time to answer questions, make suggestions, review chapters, and donate photographs during the revision of this book. I am particularly indebted to:

◆ Gianfranco Ferré who graciously gave his permission to use his wonderful sketch on the cover;

◆ Jessica McClintock whose talent, genuine interest, and cooperation in the training of fashion students makes her foreword especially meaningful.

◆ Karl Lagerfeld; Ann Wade at Hermès, Paris; Anne Etot at I. W. S., Paris; Marianne Millies-Lacroix at Galaries Lafayette; Claude Bourdier-Neff; Lauré du Pavillon at Christian Lacroix; Marie-Odile Bouillon at Peclers; and Nancy Richards in London for generously sharing exceptional information and photographs on European fashion; Heather Stuart, Beate Bowron, Laurie Belzak and Garth McGeary for information on Canada;

◆ Designers Nicole Miller, Robert Abajian at Liz Claiborne, Jeanne Allen and Marc Grant; and Odile Laugier at Adrienne Vittadini for their distinguished views on design;

◆ Robert LaForce, formerly at Hoechst-Celanese; Gene Bekaert, Greenwood Mills; Karen Puffer, Lenzing; Don Vidler, Milliken; Ray Kromer and Joan Eterton at Hoechst Celanese; John Eckert and Lynn Scott Menz at the Wool Bureau; Berrye Worsham and Bill Daddi at Cotton Inc.; and Debra Hoagland, Burlington for sharing their wealth of knowledge on textiles;

◆ Jay Margolis, formerly at Liz Claiborne; Gail Cook at Dana Buchman; Jessica Mitchell and Harriet Mosson at Liz Claiborne; and Sheila Bernstein at Adrienne Vittadini for expert advice on marketing.

◆ Don Fisher, The Gap; Sally Frame Kasaks, Ann Taylor; Bob Connolly, Wal-Mart; Walter Loeb, Loeb Associates; Julia Ellard, I. Magnin; Molly Carrara, Chanel; and Madeline Rogers, Georgiou;, for generously sharing their superior knowledge of retailing;

◆ Ian Wright and Carol Hochman at Liz Claiborne; Theresa Petry, ABS Sportswear; Casey Bush, The Millinery Information Bureau; and Ysabel Trujillo, the Canadian Fur Assn., for extensive information on accessories and furs.

◆ Ann Stock, The White House; Donna Cristina, Dente/Cristina; Rob Corder, Carolyn Galliani, Linda Fargo, and Mary Richerson at I. Magnin; and Robin Lauritano, Bloomingdale's, for their expertise in fashion promotion;

◆ Dee Katson, Levi Strauss, and Phillip Lighty, Danube Knitwear for information on imports; Jane Cook and Lillian Jang at Esprit for advice on production;

◆ Stella Warnick, Seattle Pacific University; Carol Block, Fashion Institute of Design and Merchandising; Ellen Sideri, The Fashion Works; and Jeffrey Graff, Moss Adams, for their excellent suggestions; Chris Gilbert, The Fashion Service; Linda Greenman, I. Magnin; Carol Mestrich, Macy's; Don Arruda and Mary Altshul, Gerber Garment Technology; Fern Mallis, CFDA; Barbara Colvin Hoopes, Koret; Suzie Click Lewis; and Cathleen Mackey for their expertise in various areas;

◆ my grandmother Ida Martin, my parents Ida and Russell Stephens, Elma McCarraher Page, Dolores Quinn, Eleanor Kling Ensign, Hazel Stroth, Krestine Corbin, Debra Smith, and friends and family for encouragement throughout my education and career; and

◆ most of all to my husband Philipp and my chidren Peter and Victoria for their patience.

Part One

THE FUNDAMENTALS OF FASHION

Three mannequins wearing gowns from the 1830s in a stable scene at the Costume Galleries, Castle Howard, York, England. *(Courtesy of Richard A. Robson, Curator)*

1

FASHION DEVELOPMENT

Fashion executives at every level of the industry want and need to know how the fashion business developed. Lessons in history help them to make decisions for today and the future. Ideas from the past are often reinterpreted for today's fashion.

CHAPTER OBJECTIVES

After reading this chapter you should have attained competence in the following areas:

1. Awareness of the major changes in American life styles since the Industrial Revolution and how they have influenced fashion
2. Understanding of how fashion has reflected the social, cultural, political, economic, andtechnological changes since the Industrial Revolution
3. Ability to outline and discuss major changes in the fashion industry
4. Knowledge of the names of major designers of the past 100 years

More than just a designer's whim, fashion is a subtle reflection of the social, political, economic, and artistic forces of any given time. The changing styles that evolve from these forces tell of historical events as poignantly as textbooks, journals, or periodicals. Dressing-room mirrors throughout the ages have reflected the trends in how people think, live, and love.

We will examine some of the major influences on fashion in history as a background to understanding contemporary fashion and anticipating future change. This chapter traces the development of the fashion industry in Europe and America from the seventeenth century to the present, emphasizing the last 100 years. It briefly discusses how fashion innovators, together with society, technology, economics, and politics, change fashion.

Fashion, as we know it, is relatively new. In ancient and medieval times, clothing styles remained practically unchanged for a century at a time. Fashion change began to accelerate during the Renaissance, as Western civilization discovered different cultures, customs, and costumes. As new fabrics and ideas became available, people craved more new things.

FRANCE, THE CENTER OF FASHION

France's dominance over international fashion began in the early eighteenth century.

Fashion Dictated by Royalty

Until the Industrial Revolution, people belonged to one of two main classes: the wealthy, mostly landholding aristocrats; and the poor, mostly laborers and farmers. Because wealth was concentrated in the landowning class, these people were the only ones who could afford to wear what was considered to be fashionable. Royalty, at the top of both the social and the economic ladders, set fashion trends; other members of the aristocracy followed their example in order to gain approval.

At the turn of the eighteenth century, members of King Louis XIV's court became the arbiters of taste, making Paris the fashion capital of Europe. The textile industry grew in Lyons and other French cities, supplying the court with silk fabrics, ribbons, and laces. Dressmakers and tailors, sponsored by the wealthy, developed their skills to a high level on these beautiful materials.

Hand Sewing by Dressmakers and Tailors

The elaborate detail and intricate seaming of fashion at this time required an enormous amount of painstaking hand labor. All clothes were not only handmade but also *custom-made*. Each garment was made to fit the customer's exact measurements. Dresses and suits were individually sewn by dressmakers or tailors to their employers' specifications. The identities of personal dressmakers were secrets guarded by the wealthy. No one wanted to share the talents of clever dressmakers for fear of losing them. Rose Bertin (Ber-tan´) was dressmaker to Queen Marie Antoinette, who made her the official court minister of fashion.

Poorer people wore castoff clothing from the rich or trade classes. Country folk made their own clothing. The very elaborate clothing for special occasions was passed from one generation to another and became the traditional *folk costume*. The contrast between the plight of the poor and the extravagancies of the court during the eighteenth century was one cause of the French Revolution, which began in 1789. In response to a general revulsion against excess, fashion changed from elaborately decorated costumes to simpler garments.

Worth gown typical of the Second Empire period, 1852 - 1860. *(Courtesy of the Union Française des Arts du Costume)*

Growth of the Couture

In France, the art of dressmaking was known as *couture* (koo-tour´). A male designer was a *couturier* (koo-tu-ree-ay´); his female counterpart was a *couturière* (koo-tu-ree-air´).

Charles Worth is considered the father of the couture because he was the first successful independent designer. Born in England, he came to Paris at age twenty in 1846 (the year Elias Howe patented his sewing machine). Worth attracted prominent women as clients, culminating in Empress Eugénie, wife of Emperor Napoleon III (who marked the end of royal fashion leadership).

Some couturiers became business as well as creative forces, directing salons staffed with seamstresses and tailors. Other couture houses followed Worth including Paquin, Cheruit, Doucet, Redfern, the Callot Sisters, and Jeanne Lanvin. The couture became a bridge between the class-structured fashion of the past and the democratized fashion of today.

From these beginnings, the international market for Parisian high fashion grew. In 1868 the couturiers of Paris formed a trade association. Other European capitals followed the leadership of Paris, Vienna becoming the next in importance. Couturiers were the major influence on fashion design for over one hundred years, setting style trends for all of Europe as well as the rest of the Western world.

Fashion dolls, dressed in miniature versions of couture gowns, were sent from France as a convenient means of publicizing fashion. Orders from wealthy women were mailed back to Paris, where the gowns were made to fit the customers' requirements. Some of these dolls made their way to the United States. Most people could not afford couture clothes, however, but managed to copy them to some degree.

EFFECTS OF THE INDUSTRIAL REVOLUTION ON FASHION

The Industrial Revolution marked the beginning of technological advances in textile and apparel production.

Growth of the Middle Class

Great economic, social, and fashion changes throughout the Western world accompanied the Industrial Revolution in the late eighteenth century. Burgeoning trade and industry in turn created a middle class with money to spend on the luxuries of life, including better clothing. Money gave the new middle class power, not only in business and society, but also to influence fashion trends. Fashion became a status symbol, a visual means to show off wealth.

Establishment of the Business Suit

Until 1800, men's and women's fashions had equal amounts of decoration. In Louis XIV's time, men's dress was at least as elaborate as women's. As the middle class grew, businessmen wanted to establish an image of respectability and dependability. At that point, "men's garb descended from brilliant finery . . . into bleak conformity." Men adopted the conservative, dignified business suit with long trousers, jacket, vest, shirt, and necktie—"a permanent noose, you might say."[1] Men's business attire has remained basically conservative with very few changes since.

Men's clothing, as well as women's, was custom-made. The finest tailor shops—such as Henry Poole and Company, established in 1843—were on Savile Row in London, which became the international center of men's fashion.

Some ready-made men's clothing was made by hand in France in the late 1700s. In America, the first ready-to-wear suits were made for sailors so that they would have clothes to wear when they came on land. Naturally, these first clothing factories were located in seaport cities, such as New Bedford, Boston, New York, and Philadelphia. The first record of a men's clothing factory is dated 1825.[2]

In 1818, Henry Brooks started the men's clothier business that became Brooks Brothers in 1850. There, Abraham Lincoln bought an overcoat for his second inauguration.[3] In 1824, Jacob Reed opened his first store in Philadelphia.

Growth of the Textile Industry

Early America had virtually no textile or fashion industry. Most materials were imported from abroad: silks from Italy, France, India, and China; and woolens, calicoes, and cashmeres from Britain.

The modern textile industry, which enabled more fabrics to be produced in less time, began in England with John Kay's development of the flying shuttle in 1733, James Hargreaves's invention of the spinning jenny in 1764, Richard Arkwright's water frame in 1769, and Edmund Cartwright's power loom in 1785. To protect its industry, England passed strict laws preventing textile machines, parts, blueprints, tools, and even the mechanics and inventors themselves from leaving the country. However, Samuel Slater memorized every detail of Arkwright's water frame and other machinery and secretly left England. Within two years of his arrival in New England, he had a new mill built and in operation. Textile mills began to produce cloth in America, the first evidence of fashion independence. New England became America's first textile center.

In the late 1800s, the American textile industry began to relocate to the South, the source of cotton. Southern states continued to offer other incentives such as cheaper labor costs. Eventually the South became the center of the textile industry in the United States.

MASS PRODUCTION OF CLOTHING

The mass production of clothing led to accessible fashion for everyone.

Invention of the Sewing Machine

The democratization of fashion began with the sewing machine, which turned a handicraft into an industry. The mass production of clothing would have been impossible without it, and without mass production fashion would not be available to everyone. In 1829, a French tailor named Thimmonier patented a wooden chain-stitch sewing machine, but all

The first Singer sewing machine, 1851. *(Courtesy of the Singer Company)*

existing models were later destroyed by rioting tailors who feared for their jobs. Walter Hunt, an American, developed a sewing machine in 1832 but failed to patent it. Thus, the man who is usually credited with its invention is Elias Howe, who patented his in 1846. All of Howe's machines were run by hand. In 1859, Isaac Singer, whose name has become a household word because of his mass production of the sewing machine, developed the foot treadle, an improvement that left the hands free to guide the fabric. Singer spent $1 million a year on sales promotion, and by 1867 was producing a thousand machines per day.[4] Electrically powered models were not available until 1921.

An early use for sewing machines was to make military uniforms, but soon they were being used for the mass production of everyday men's wear. At first, manufacturing was done by the *cottage industry* process. Garments were cut and bundled in a plant and sent out to homes for sewing. Later, to save time and the cost of delivering pieces and picking up garments, and to keep control over production, entrepreneurs brought workers and machinery together in factories. This caused many people in search of work to move to the cities where the factories were located.

Women's Fashion Reflects Social Changes

Fashion conveyed the rigid differences between the roles of the sexes. Men wore trousers, which became a symbol of dominance, while women wore constraining garments characteristic of their restricted life styles and obedience to their husbands and fathers.

Since the development of practical, simple clothes for men, fashion has usually emphasized women's wear. As a result, women have often been criticized for paying too much attention to it. That interest seems perfectly understandable, however, when we consider that until recently men did not give women the right to own anything but their wardrobes.[5]

Aside from the small number of wealthy women who bought couture, most women had about three basic garments in their wardrobes. Even after the invention of the sewing machine, only hoop skirts and cloaks could be manufactured for women. Fashionable one-piece fitted dresses were impossible to mass-produce because each dress had to be custom-made to fit at least three sets of usually irregular measurements.

Mass Production of Women's Separates

The introduction of separate blouses and skirts in the 1880s made it possible to manufacture ready-to-wear clothes for women. A blouse could

A young woman dressed in the Gibson Girl style.
(Courtesy of the National Archives, Washington, D.C.)

be made to fit the shoulder and bust measurements, the skirt to fit the hips. Waistlines and hemlines were easily adjusted and blouses were simply tucked in. This innovation made it possible for the working- or middle-class woman to add variety to her wardrobe simply by mixing separates. The cost of a new ready-made blouse was a mere fraction of the cost of a custom-tailored dress.

The *Gibson Girl* was the personification of the ideal young middle-class American woman, as sketched by popular illustrator Charles Dana Gibson in the 1890s. The Gibson Girl gave style to the basic high-necked, long-sleeved blouse-and-skirt look. It was practical yet feminine and could be worn anywhere. By 1900, the American labor force included more than 5.3 million women at all levels of the social scale, properly attired in practical blouses and skirts.

Besides some businessmen's suits and working women's blouses and skirts, the only other manufactured clothes were uniforms or workingmen's gear. Levi Strauss started making his sturdy denim pants for miners in 1850. Both Levis and the Gibson Girl look paved the way for the simplified, functional dress that typifies American fashion.

Children's Fashion

The wealthy were the only ones who had money to spend on fashionable children's clothes; members of the middle and working classes made their children's clothes at home. Small children, both girls and boys, wore dresses. As they grew older, children were supposed to act like adults and they were dressed in miniature versions of adult apparel. In fact, many children wore cut-down remakes of their parents' old clothes. Mothers were particularly grateful for the advent of patterns for children's clothes because previously, home-sewn garments had been cut and fitted by trial and error.

Paper patterns, inspired by French fashions, were made available to American home sewers in 1850 by Ellen and William Demorest.[6] Demorest Patterns, followed by Butterick and McCall's, fostered fashion consciousness at all levels of society. Women on small budgets were especially happy to have patterns to make the clothes they could never afford to buy.

RETAILING DURING THE NINETEENTH CENTURY

Retail executives, including merchandisers and managers, learn about the growth of retailing to make informed decisions on retailing today and for the future.

Fairs and bazaars were the predecessors of the retail store. The traveling merchant brought clothes to these markets. Expensive goods were shown only to selected wealthy customers. Prices were not marked on the merchandise, so buyer and seller usually bargained.

As large numbers of people settled in towns, the first general stores were established to cater to their desire for wider assortments of merchandise. Also, artisans sold their handmade goods in their own shops. These shops were grouped together by trade and regulated by guilds.

The Industrial Revolution triggered a self-supporting manufacturing and retailing cycle. As more goods were produced, there were more products to sell. This increased business activity gave the growing middle class more money to spend, which created a demand for more products. This growing demand for the variety of goods being produced was the basis for the growth of retailing. Retail stores grew up in the cities, close to production and population centers. As more people clustered in cities to work, stores opened in areas convenient to shoppers.

Two types of stores finally emerged to bring fashion to the public: the *specialty* store and the *department* store. Traditional handicraft stores evolved into specialty stores, which grew in importance in the late nineteenth and early twentieth centuries. An outgrowth of the general store, the department store which carried a wide variety of merchandise, was a product of the nineteenth century. Shopping in department stores became a popular activity, like going to an exhibition. For the first time, people of all incomes could at least enjoy browsing and looking at beautiful things.

The First Department Stores

In 1826, Samuel Lord and George Washington Taylor formed a partnership to open the first Lord and Taylor store in New York City. Jordan Marsh and Company, opening in Boston, claimed they could sell, cut, sew, trim, and furnish a dress in half a day.[7] Edward Filene opened a

John Wanamaker's Grand Depot store, Philadelphia, 1877, located on the same site as today's downtown store.

comparable department store in Boston; John Wanamaker in Philadelphia; Joseph Hudson in Detroit; Morris Rich in Atlanta; and R. H. Macy, and Nathan Straus in New York.

The American contribution to retailing is not only size but also customer service. In Chicago, Marshall Field once admonished a store clerk who was arguing with a customer, "Give the lady what she wants." "The customer is always right" has been a principle of American retailing ever since.

Harrod's of London, established by Henry Harrod in 1849, has become the largest department store in Europe. It began as a small grocery store, and by 1880 it had 100 employees. Liberty of London opened its retail store in 1875 and produced its own prints as early as 1878. In France, department stores such as the Bon Marché and Printemps opened in the nineteenth century.

Early Mail-Order Merchandising

In the 1800s, nearly three-quarters of the American population lived in rural areas, usually served only by a few general stores with a limited selection. The extension of the railroads to the West Coast and the inauguration of a free rural mail delivery enabled merchants to start reaching these potential consumers with mail-order services.

While working for a wholesaling firm, traveling to country stores by horse and buggy, Aaron Montgomery Ward conceived the idea of selling directly to country people by mail. He opened his business in 1872 with a one-page list of items that cost one dollar each. People could later order goods through a distributed catalog, and the store would ship the merchandise cash on delivery (COD). The idea was slow to catch on because people were suspicious of a strange name. However, in 1875 Ward announced the startling policy of "satisfaction guaranteed or your money back." Contrasting with the former retailing principle of *caveat emptor* (Latin for "buyer beware"), this policy set off a boom in Ward's business.

In 1886, a Chicago jewelry company erroneously shipped some watches to a jeweler in Richard Sears' hometown in Minnesota. Sears offered to resell them for the jeweler, thereby creating his own watch business. Alvah Roebuck answered Sears' ad for a watchmaker and became Sears' business partner. In 1893, the firm name was changed to Sears, Roebuck and Company. From a modest beginning, they expanded by 1895 to a 507-page catalog including clothing and household goods, often referred to as the "dream book" or "wish book." The mail-order business did more to bring a variety of up-to-date merchandise to rural consumers than any other form of retailing.

CHANGES CAUSED BY COMMUNICATIONS, LEISURE, AND INDUSTRY

Communications, leisure activities, labor conditions, and industrial technology have a continuing effect on fashion.

The desire for fashionable clothing was fostered by its increased availability as well as by new communications media such as the mail service, magazines, newspapers, telephones, automobiles, airplane travel, and later radio, motion pictures, and television.

The First Fashion Magazines

During the 1800s fashion magazines began to be published in France and England. Eighteen fashion magazines were being published in New York and Philadelphia in the late 1800s.[8] Two magazines that began in the nineteenth century are still published today. An outgrowth of *Harper's,* modeled after *Der Bazar* magazine of Berlin, *Harper's Bazaar* commenced publication in both New York and Paris in 1867. *Vogue* started in 1894 in New York.

These publications spread the latest fashion ideas from Paris by means of sketches and descriptions. Dressmakers in other countries copied the styles as best they could with available fabrics. As more women became aware of fashion styles through magazines and other forms of mass communication, their desire to wear those fashions increased. The faster a style was adopted by the public, the greater was the demand for more new looks.

The Growth of Leisure Activities

The popularity of sports such as tennis and bicycling created a need for functional sportswear. As early as 1851 Amelia Jenks Bloomer had tried to introduce pants for women. However, they were not accepted until the bicycling craze of the 1890s. Bloomers, full in the leg and gathered at the ankle, were also worn under bathing dresses. Finally, after 1900, swimwear was pared down enough that people could actually swim.

Pants became an acceptable part of the horseriding habit for women around the turn of the century, when women discovered that riding sidesaddle in a skirt, "often caused one to dismount before the ride was over."[9] As women became more and more involved in sports, pants gave them the mobility for a more active life. However, it was not until the 1920s that pants became fashionable as well as functional for women.

Conditions in the Garment Industry

New York Becomes the Center of the U.S. Fashion Industry

Until 1850 the garment business in the United States was concentrated in Boston, Philadelphia, and Baltimore. By the latter part of the century, however, the influx of European immigrants to New York had helped to establish that city as the center of the industry. The immigrants, used to hardship and willing to work for low wages, provided the skilled labor the industry needed in order to grow. By 1900 the American women's clothing industry consisted of 2701 establishments.[10] They produced mostly cloaks and suits, with some shirtwaists (blouses) and underwear.

Women working in Levi Strauss' Valencia Street factory.
(Courtesy of Levi Strauss & Co.)

Unionization

As more workers crowded into the industry, working conditions became appalling. Tenement workrooms were known as *sweatshops* because of the excessively long hours required of laborers in unsanitary surroundings for extremely low wages.

In 1900 cloak makers, mostly immigrants living in cities in the northern United States, met to discuss working conditions. The result was the formation of the International Ladies' Garment Workers' Union, which tried to protect its members against unfair employers . At first the union was not very popular, but it did make progress with strikes against the shirtwaist industry in 1909 and the cloak industry in 1910. "On March 25, 1911, the nation was stunned by the horror of the Triangle Shirtwaist Company fire in New York City."[11] The factory's main exit door had been bolted, and the lone fire escape was a death trap that ended in midair. The 146 deaths, mostly of girls, aroused Americans' indignation against the plight of the sweatshop workers. Finally, action was taken on demands for regular hours, minimum wages, paid vacations, sick benefits, and better working conditions. Added labor costs naturally added to the inevitable simplification of fashion.

EFFECTS OF WORLD WAR I ON THE STATUS OF WOMEN AND FASHION

Women in the Work Force

The status of women has hinged firmly on their participation in the working world. Without a prominent place in business, women had no authority and no rights. Male dominance was hard to overcome, and the old patterns did not begin to change until the period around World War I. In 1914 the war began in Europe, and the United States entered in 1917. World War I greatly promoted women's rights, because it enabled European and American women to replace men in previously all-male jobs. The functional working clothes worn by these women had a great impact on fashion. "Now that women work," *Vogue* reported in 1918, "working clothes have acquired a new social status and a new chic."[12]

Clothes Tailored to Jobs

Fashion change reflected the emancipation of women. Women were freed from laced corsets because of their participation in the work force. No one wanted or had time for complicated dressing. This change coincided with the need to simplify clothing construction because of rising labor costs and resulted in the democratization of fashion.

Women's clothing of this period asserted that women were as useful as men. There was a trend toward masculinity in women's fashion: decorative details disappeared in favor of a tailored look that imitated businessmen's suits. Corsets were discarded and the curved hourglass silhouette was replaced by the tube. Hemlines rose and skirts widened to permit freedom of movement. Fashion reflected women's growing independence and, in 1920, women finally won the right to vote in the United States.

Important Trendsetting Designers

While mass production was growing in the American fashion industry, the French couture still concentrated on fashion leadership among the wealthy. Often one or a few designers dominate the field because they are able to capture the spirit of their times and translate it into highly accepted fashion.

Paul Poiret (Pwah-ray´), whose tubular dresses liberated women from corsets, was the first Paris couturier of this century to become a trendsetter.

Gabrielle Chanel (Sha-nelle´), also known as *Coco*, was at the forefront of French fashion following World War I. Chanel popularized the *Garçon* (gar-sohn´) or boyish style, with sweaters and jersey dresses and was the first designer to make high-fashion pants for women.

Madeleine Vionnet (Vee-ohn-ay´) originated the body-skimming, bias-cut dress.

Jean Patou (Gsahn Pa-Too´) created the famous *Flapper* look in 1925 by accentuating the hipline, strengthening a straight silhouette, and making shorter skirts with uneven hemlines. He was the first couturier to open his collection showings to the press.

The ready-to-wear apparel industry began to prosper when designers such as Poiret, Vionnet, and Chanel simplified styles and thereby construction. Couture styles were then copied by mass producers for consumers at every price level. Because individual fitting was not so important to their straight silhouettes, mass production of dresses became practical. As early as the 1920s, designers such as Lucien Lelong in France and Hattie Carnegie in America were adding ready-to-wear lines to their made-to-order collections. By the late 1920s American ready-to-wear was firmly established.

Coco Chanel, wearing one of the suits she made famous. *(Courtesy of Chanel, Paris, photographed by Hatami.)*

Jean Patou and his American models arriving at Le Havre in 1924. *(Courtesy of the National Archives, Washington, D.C.)*

RETAIL EXPANSION IN THE EARLY TWENTIETH CENTURY

The growing middle class had more money to buy, which created greater production and the need for more stores.

Specialty Stores for Quality Fashion

Specialty stores tried new retailing approaches and offered their customers high-fashion merchandise. Bergdorf Goodman and Saks Fifth Avenue in New York City and Neiman Marcus in Dallas concentrated solely on the finest fashion and customer service. By the 1930s the first women presidents of major retail firms had been installed, Dorothy Shaver at Lord and Taylor and Hortense Odlam at Bonwit Teller. Shaver gave American fashion a boost by mentioning American designers in store ads.

The Expansion of Chain Stores

While great retailing establishments were growing in the big cities, chain stores selling lower-priced merchandise were taking hold elsewhere. James Cash Penney was such an industrious employee at a small Wyoming store that the owners offered him a partnership in their new store in 1902. Called the Golden Rule Store in honor of their belief in high business standards, it proved an immediate success, in part because of Penney's door-to-door advertising campaign. In 1907 the original partners sold their shares to Penney; the store's name was officially changed to J. C. Penney in 1912. When the chain-store concept caught on in the 1920s, Penney opened stores in all parts of the United States.

Joan Crawford glances admiringly at Adrian, Hollywood trendsetter of the 1930s. *(Courtesy of the Joseph Simms Collection)*

The Advent of Suburban Shopping Centers

As more and more people owned cars, personal mobility increased, creating a revolution in retailing. Finding that its mail-order business was dropping off, Sears & Roebuck opened stores not in the city centers but rather near the highways that led to the growing suburbs, where they could offer free parking. It was also the beginning of suburban shopping centers such as The Country Club Plaza in Kansas City, which opened in 1922.

EFFECTS OF THE DEPRESSION ON FASHION

The experience of the Great Depression of the 1930s still causes manufacturers and retailers to worry at the sign of a recession.

The Bursting of the Credit Bubble

In the 1920s, so much credit was extended that eventually there was not enough money to back it up. In the stock market a person needed to put up only 10 percent of the price to buy stock; when the price rose, the shares could be sold at a profit. So it went until September 3, 1929, when the stock market reached an unbelievable height. Then it started a steep decline. In less than a month the market value of all stocks dropped $30 billion. Unemployment rose from 1.5 million to 12.8 million, and business profits fell from $10.3 billion to a net loss of $2 billion. Nearly half of the nation's banks had to close. Industrial production fell to half of what it had been and many companies went bankrupt. More than a third of the ready-to-wear manufacturers went out of business. The slump set off a chain reaction that soon put the whole world into a depression.

Hollywood's Influence on Fashion

Americans tried to take their minds off the Depression at the movies. Because people in pre-television days commonly visited the local movie theater once or twice a week, American films brought fashion to every woman. Every young woman wanted to look like her favorite film star. Katharine Hepburn and Marlene Dietrich made slacks popular for women; Clark Gable popularized the sport shirt for men. The 1930s were the most glamorous years in film history, a paradoxical contrast with the deprivations of real life.

Gilbert Adrian emerged as the leading Hollywood designer. He was the first American designer to influence fashion throughout the world. He designed a ruffled dress with puffed sleeves for Joan Crawford to wear in *Letty Lynton* in 1932. As an example of Adrian's impact on ready-to-wear, Macy's alone subsequently sold half a million simplified versions of that dress at affordable prices.[13] The dress was made of rayon, then a new synthetic fiber created to imitate silk.

Paris' Influence on International Fashion

Elsa Schiaparelli (ska-pa-rell´-ee) was the trendsetter of European fashion in the 1930s. "Schiap" got her start in Paris with a large white bow motif knitted into a sweater front, the forerunner of our modern sweater designs and of the decorated T-shirt. Schiaparelli used bold accents of color, especially shocking pink, which she made famous. Schiaparelli moved the center of interest to the shoulders, which she began to widen, accentuating them by pleats, padding, or braid, a silhouette that remained popular through World War II.

James Mainbocher (Main-bow-shay´) was the first American designer to be successful in Europe. Combining his middle name Main and last name Bocher into Mainbocher, he opened a salon in Paris in 1930. In 1937 King Edward VIII of England created a scandal by abdicating his throne to marry the divorced American Wallis Simpson. Due to extensive publicity, her luxurious life style, and her sensitivity to fashion, the future Duchess of Windsor was regarded as one of the best dressed women of her time. Her selection of a wedding dress designed by Mainbocher, which became the most copied dress of the 1930s, made him a trendsetter.

Gown by Schiaparelli. (*Courtesy of Claudy Stolz and Maison Schiaparelli, Paris*)

WORLD WAR II'S EFFECT ON FASHION

The American economy did not entirely recover until World War II escalated production. Retailers and manufacturers are not the only people who would like to avoid the need for this in the future.

America's Isolation from Paris Fashion

During the war the French couture banded together under the leadership of Lucien Lelong, then president of the Paris Couture Syndicale, in order to avoid being sent to Berlin due to the German occupation. Under great restrictions and privation—practically no fabrics to work with, no trimmings, no press coverage, no heat, and little food—most designers barely managed to stay in business. Some were forced to close. Of course, under these circumstances little was achieved.

Isolated from Paris fashion leadership during the war, Americans had to find their own style direction. The lack of imports from France was actually a boon to the development of American talent. In 1940 *Vogue* reported on the New York collection openings. With Mainbocher as an example of success, other American designers such as Claire McCardell, Hattie Carnegie, and Vera Maxwell gained recognition.

Claire McCardell, considered the top American designer, was credited with originating the *American Look* in practical separates, inspired by the work clothes of farmers, railroad engineers, soldiers, and sportsmen. American designers became especially skilled at and known for their sportswear, reflecting the more casual American life style, which would eventually influence the rest of the world. Sportswear, with its simpler construction, also suited mass production.

Fashion remained relatively stable during the war years. Women doing war work wore uniforms or work clothes so that functional clothes became a necessity. The U. S. government's wartime regulations restricted the use of fabric and hardware. The result was a masculine silhouette for the women who now shouldered the responsibilities at home.

Claire McCardell sketching in a museum.
(Courtesy of the National Archives, Washington, D.C.)

POSTWAR FASHION

Women were tired of the clothes made under wartime restrictions; they were ready for change.

Emphasis on Femininity

Women were so happy to see men home alive after the war that most reverted completely to stereotypical feminine roles, leaving jobs open for the returning men. Fashion catered to their feminine ideal.

Christian Dior (Chris´-tee-ahn´ Dee-or´) showed his first collection in 1947 and was an instant success. In a reaction against the wartime silhouette, women adopted his *New Look,* with longer, fuller skirts, and smooth, rounded, sloping shoulders and tiny fitted waists. Within a few seasons Dior's name became a household word and he was doing as much business as the rest of the couture combined.[14] Paris recaptured fashion dominance—almost to the point of dictatorship in Dior's case.

Christobal Balenciaga (Bah-lehn´-see-ah´-gah), a Spaniard who worked in Paris, was regarded as the master of tailors. When American stores purchased rights to manufacture *line-for-line* copies, Balenciaga's designs were always the most popular.

Dior's New Look. *(Courtesy of the House of Dior)*

American Fashion Innovators of the Postwar Period

While Parisian designers set international trends, Americans also enjoyed continued success at home due to the exposure given to domestic fashion during the war. These designers included: Bonnie Cashin, Oleg Cassini, Ann Fogarty, James Galanos, Charles James, Anne Klein, Norman Norell, Mollie Parnis, Fernando Sarmi, Adele Simpson, Jacques Tiffeau, Pauline Trigere, Sydney Wragge, and Ben Zuckerman. Don Loper and John Weitz developed the first coordinated sportswear for men.

Some American designers custom-made fashion for the wealthy, but most built their reputation on what Americans did best, ready-to-wear. The French may have been the high fashion innovators but Americans developed and excelled at producing fashion looks for everyone.

Jacqueline Kennedy, wife of the president, set a new ideal of beauty and established an impeccable look predicated on simplicity that was universally imitated. Jackie practically made a uniform of her two-piece jewel-necked A-line dresses and pillbox hats. However, the late 1950s and early 1960s saw the last of elegant fashion domination for twenty years.

Accent on Family Life

In the postwar search for domestic tranquility, American families wanted to escape the deteriorating cities and find a healthy environment in which to raise their children. For many this meant a move to the suburbs. The casual suburban lifestyle brought about the popularity of casual sportswear by American designers, new wash-and-wear manufactured fabrics, and more convenient shopping centers.

The late 1940s and 1950s saw the beginning of the biggest increase of births ever. As these children became teenagers, industry catered to this newly emerging market. Records, cosmetics, magazines, and "junior" fashions were created.

THE NINETEEN SIXTIES

The postwar baby boom had an increasing effect on fashion change. Breaking with convention, young designers created fashions for their own age group.

By 1965, 50 percent of the United States population was under the age of 25. Sheer numbers brought increased buying power and encouraged a youth-oriented market. Young people's tastes were to dominate the fashion scene through the 1970s.

London Emerges as a Leader in Youthful Fashion

Mary Quant, and other young British designers such as Zandra Rhodes and Jean Muir, set international fashion trends. They were influenced by a group called the "Mods" who put together odd separates and old clothes from flea markets to create an individual look. Miniskirts, which rose above the knee, tights (panty hose) and unusual fabrications such as vinyl, were characteristic of the *Mod* look.

1960s trendsetter Mary Quant.
(Courtesy of Mary Quant Limited)

In the United States young designers such as Betsey Johnson also created youthful fashions. Even the couture designers in Paris, such as André Courrèges, followed the lead of these young designers. This was the first evidence of a reverse in the traditional fashion-adoption process. The popularity of the youthful look made all women want to look young.

Revival of Men's Fashion

The English Mod look affected men's wear as well as women's. Carnaby Street tailors made a brave attempt to return color and fashion to men's clothing. The initial impact of Carnaby Street did not last, but the general awareness of the need for more interesting men's clothes did. Men became more concerned with their roles outside of work and with dressing for those other sides of life. French and Italian designers became as important in men's wear as the traditional English.

Pierre Cardin (car-dahn´) signed his first contract for men's shirts and ties in 1959 and opened his men's ready-to-wear department in 1961. Dior, St. Laurent, and other women's designers followed his example. The 1960s brought the first designer clothes for men and the first extensive fashion changes since the introduction of the business suit.

Boutiques Set Retailing Trends

British boutiques such as Mary Quant's Bazaar set a new trend in retailing. The French term *boutique* was adopted by Western countries as these small shops gained popularity. Yves St. Laurent opened his Rive Gauche (reev gōsh) boutiques around the world. Henri Bendel's in New York introduced an atmosphere of many boutiques within one store. This idea brought freshness and excitement to retailing.

THE NINETEEN SEVENTIES

Antifashion became the style statement of the Seventies.

Antifashion

The tumultuous late 1960s—a time of assassinations, riots, and civil strife—made people turn away from showy displays of frivolity. In suburban America the polyester pantsuit became almost a uniform. It even became stylish to look poor. Workmen's Levis became an antifashion statement and the importance of denim remains today. Some young people, bored with the lack of excitement in retail merchandise, wore vintage clothing purchased at thrift stores.

The Ethnic Look

Society's dropouts, the "hippies," had an influence on fashion in their tattered jeans, long hair, beads, and old clothes. In their desire to emulate the simple life, they combined their old clothes to simulate ethnic or native American costume.

Eventually the *Ethnic* look widened to include traditional folk costumes from practically every country: India, Guatemala, Greece, and the African nations. Black people developed pride in their heritage, sporting Afro hairdos and the dashiki, traditional African garb. The resumption of United States relations with China in 1972 created widespread interest in the Chinese people and Chinese costume. Designers incorporated into their collections the ethnic idea of layering and combining separates.

Yves St. Laurent (Eve Sahn´Law-rahn´) emerged as the fashion star of the Seventies because he was able to interpret ethnic and other street looks into high fashion. St. Laurent's Rive Gauche boutiques featured his prêt-à-porter (ready-to-wear) collection, which signaled the end of couture dominance. At that time, he was best known for his blazers, city pants, and ethnic looks. The epitome of reverse fashion adoption was St. Laurent's Elegant Peasant collection of 1975: ethnic looks done in silks, with high price tags.

1970s trendsetter Yves Saint Laurent and a gabardine pantsuit from his 1969 collection. *(Courtesy of Yves Saint Laurent)*

Physical Fitness as Fashion

In the 1970s people became more aware that beauty was more than skin deep and set about to keep their bodies fit with exercise. Jogging was very popular along with traditional sports such as tennis. Fashion was not lost in the race, and soon everyone had a jogging suit, even if they did not run. By the 1980s, active sportswear was a firmly established fashion category. Designers and manufacturers created clothing for every sport including cycling. The elastic fiber spandex was developed to give garments the stretch needed for movement.

The denim look of the 1970s culminated in "designer jeans" by Calvin Klein and Gloria Vanderbilt. This was followed by a wave of conservative, nostalgic leisure dressing in khaki shorts, penny loafers, Izod Lacoste shirts, button-down shirts, and blazers for men and women, called the *Preppy* look.

The Women's Movement

The 1970s witnessed women struggling to find an equal place in the business world. Women were trying to make it up the corporate ladder rather than the social one. In order to fit into a man's world, they adopted the conservative business suit (with skirt) to give themselves a visual businesslike credibility. "Dressing for success" became the byword for those who wanted to get ahead. In a somewhat misguided attempt at quality consciousness, status dressing became important. The label of the garment became more important than the design. Calvin Klein, Halston, and Mary McFadden became important designer names. Rolex watches, Gucci shoes, and Louis Vuitton bags became status accessories.

THE NINETEEN EIGHTIES

Overspending and overborrowing in the eighties caused many of the problems that the fashion business faces today.

Global Fashion

During the 1980s, fashion became a global phenomenon. Italy became a major international fashion capital; Missoni knits, Armani suits, and Krizia sportswear were in great demand worldwide.

Giorgio Armani, of Milan, Italy, became the trendsetter of the 1980s because his tailored look so perfectly suited the career woman and he remained a major influence. His $200 million empire includes Emporio Armani (stores and a less expensive collection) and licenses.

Japan emerged as an important fashion capital in the early 1980s led by Kenzo Takada and then Issey Miyake. Japanese designers showed their collections in Paris and influenced world fashion with their oversized silhouettes, wrapping, and layering.

1980s trendsetter Giorgio Armani.
(Courtesy of Giorgio Armani SpA)

In 1981, the eyes of the world were on England where Prince Charles wed Lady Diana Spencer. Copies of her wedding gown were in London shop windows eight hours after the wedding!

In France, Karl Lagerfeld helped to bring elegance back to fashion and Christian Lacroix's flamboyant silhouettes and colors influenced international evening fashion.

The 1980s also saw the beginning of international recognition of American designers, led by Calvin Klein. This prompted other American fashion companies to consider exporting. Ralph Lauren gave the *American* look elegance while Liz Claiborne brought fashion to middle price ranges and the growth of both of their companies was astounding. The success of these designers was stimulated by a new appreciation of quality and luxury as Americans went on a spending binge.

Industry Trends

Textile and apparel manufacturers, faced with increased competition from imports, turned to importing themselves. In an effort to compete against imports, the fashion industry began the *Quick Response* program, involving the use of computers and electronic data interchange to foster cooperation between textile and apparel producers and retailers.

Retailing went through an optimistic period which unfortunately caused overexpansion. We became an *overstored* (industry jargon for too many stores) nation besieged with leveraged buyouts and takeovers. Mail-order retailing grew tremendously, attracting busy career women who preferred to shop by catalog. Nordstrom, a Seattle-based retailer, built a national reputation for customer service that set a standard for other stores to follow.

THE NINETEEN NINETIES

In the last decade of the century, we've had to readjust to a less indulgent way of life.

Recession

The biggest impact of the early 1990s was an international recession beginning in the United States and the United Kingdom and finally reaching Japan and continental Europe. This had far-reaching consequences for the fashion industry.

Retail overexpansion in the eighties resulted in overwhelming competition, forcing bankruptcies and store closures in the nineties. Sears gave up its catalog, a veritable American retail institution. Specialty and department stores were forced to be more focused and value oriented. In 1991, more than half of all apparel sold in the U.S. was bought on sale.[15] Discount and off-price stores such as Wal-Mart enjoyed success in spite of the recession because their low prices appealed to consumers. Retailers are trying to win back customers with various new strategies including value, customer service, and private label collections.

Naturally apparel manufacturers were affected by the recession as there are less stores to sell to and less money spent by consumers. Textile and apparel producers joined retailers to debate the pros and cons of the North American Free Trade Agreement. Textile producers hope that manufacturers will be able to use U.S. textiles if garments are produced in Mexico instead of Asia. They also want Carribean Basin countries to be included in the agreement.

Fashion Direction

The recession added confusion to fashion with a variety of hemlengths and pants, tailored looks including mini skirts and city shorts, and softer ethnic looks influenced by the seventies.

Karl Lagerfeld, dubbed "King of Paris fashion" by the press, increased his influence and became the foremost international trendsetter with five major collections: Chanel, Lagerfeld, Chloe, Fendi, and KL. His designs have an impact on many market segments and prices ranges from elegant coats and suits to junior dresses.

Successful american designers such as Donna Karan, Calvin Klein and Richard Tyler dominate the domestic designer market. Their designs are

1990s international trendsetting designer Karl Lagerfeld. *(Courtesy of Karl Lagerfeld and Chanel)*

appropriate for American life styles, are favored by the fashion press, and are copied at lower price levels.

Fashion today no longer follows one direction. Many segments of the market have been created by various life styles and tastes. Depending on age, position, mood, interests, and income, people want many more acceptable fashion choices at reasonable prices. Retailers and manufacturers alike face a challenge in the future. The turn of the century has already become a focal point for fashion.

SUMMARY

This chapter has briefly covered the growth of the fashion industry. Technological advances, especially the invention of the sewing machine, changed clothing production from custom-made to ready-to-wear. The Industrial Revolution also nourished the growth of a large middle class, who demanded and could afford fashion at every price level. As a result, fashion became available to everyone instead of just the wealthy few. Fashion has also been influenced by the changing status of women and by the changing roles of both sexes. Fashion leadership originated with and was maintained by the French, except during World War II and in the 1960s and 1970s. There are currently three major fashion capitals: Paris, New York, and Milan and there is no longer one fashion direction. After a long period of dictating fashion, the industry now tries to cater to the needs of the consumer by supplying clothes for business and leisure. Today, people demand quality and value in their clothing, which has paved the way for the success of discount and off-price stores.

CHAPTER REVIEW

Terms and Concepts

Briefly identify and discuss the following terms and concepts:

1. France, the fashion center
2. Royalty as trendsetters
3. Development of the couture
4. Growth of the middle class
5. Growth of the textile industry
6. Impact of the invention of the sewing machine
7. Development of mass production
8. The Gibson Girl
9. Development of department stores
10. Development of specialty stores
11. Development of mail-order houses
12. Fashion magazines
13. Development of active sportswear
14. Garment-industry development and conditions
15. Influence of the change in women's status
16. Retail expansion
17. Effects of World War II
18. The *New Look*
19. Postwar baby boom
20. Antifashion
21. The *Ethnic* look

Questions for Review

1. Discuss the change in emphasis over the years from couture fashions to ready-to-wear.
2. Give two examples of how fashion reflects social or political history.
3. What factors made Paris the center of world fashion?
4. Discuss the growth of mass production in the United States and the technical development that made it possible.
5. Discuss retail development and expansion in the nineteenth and twentieth centuries. How did these reflect social changes?
6. Discuss four of the most important influences on women's fashion styles between 1850 and 1920.

Projects for Additional Learning

1. Trace the fashions of one of the designers listed in the Influential Designers Chart by consulting newspapers and magazines from the period in which he or she was best known. Trace the evolution of the designer's styles with sketches or photocopies. Discuss the characteristics that made his or her designs unique. How did the designs reflect lifestyles?
2. Start a costume collection. Canvass your family, friends, and neighbors for attic donations, or shop at flea markets and the Salvation Army. Label all items carefully as to donor and year made. Examine them for construction methods and design details. Design students can use these items as inspiration for future design projects.
3. Visit a local sewing-machine dealer. Ask the dealer to trace the technical advances in machines as far back as he or she can remember. The dealer may have old catalogs or old machines that you can examine. Discuss the advantages of the old and the new models.

NOTES

[1] Phyllis Feldkamp, "Men's Fashion, 1750-1975," *New York Times Magazine,* September 14, 1975, p. 66.

[2] "Bicentennial of American Textiles," *American Fabrics and Fashions,* no. 106 (Winter-Spring 1976), 64.

[3] Ibid., p. 12.

[4] Ishbel Ross, *Crusades and Crinolines* (New York: Harper & Row, 1963), p. 12.

[5] Ibid., p. 99.

[6] Ibid., p. 20.

[7] Ibid., p. 116.

[8] Ibid., p. 220.

[9] "Women's Pants," *L'Officiel USA,* Spring 1977, p. 109.

[10] Florence S. Richards, *The Ready to Wear Industry 1900-1950* (New York: Fairchild Publications, 1951), p. 8.

[11] *Signature of 450,000* (New York: International Ladies' Garment Workers Union, 1965), p. 24.

[12] Quoted by Helen Brockman, *The Theory of Fashion Design* (New York: John Wiley, 1965), p. 69.

[13] Ernestine Carter, *The Changing World of Fashion* (London: Weidenfeld & Nicolson, 1977), p. 70.

[14] Charlotte Calasibetta, *Fairchild's Dictionary of Fashion* (New York: Fairchild Publications, 1975), p. 561.

[15] Ira Schneiderman, "A Look at the Future," *Women's Wear Daily,* August 1992, p. 15.

HISTORICAL CHART
OF INFLUENTIAL DESIGNERS

	years most influential designer and international fashion directions
1774 - 93	Rose Bertin: dressmaker to Marie Antoinette
1790 - 1815	Hippolyte le Roy: dressmaker for the court of Napoleon, creator of the classic revival Empire style
1860s	Charles Worth (b. England): father of modern couture
Late 1800s	Redfern, Cheruit, Doucet, Paquin
Early 1900s	Madame Gerber (house of Callot Sisters), Jeanne Lanvin
1909 - 11	Paul Poiret: his tunics freed women from corsets
1912 - 15	Charlotte Premet
Post-World War I	Madelaine Vionnet: first to do bias cut
1916 - 21	Coco Chanel: known for the Boyish Look and for using jersey
1922 - 29	Jean Patou: known for the Flapper Look
1930 - 35	Elsa Schiaparelli (b. Italy): hard chic and unconventional styling
1936 - 38	Mainbocher (b. United States) and Molyneux (b. Ireland): understatement and broadening shoulders (All of the above worked in France.) Gilbert Adrian: Hollywood glamour copied in ready-to-wear
1940 - 45	Claire McCardle (American): known for American Look of practical sportswear. The war years made communication with Europe impossible and made Americans begin to appreciate their own designers.
1947 - 57	Christian Dior: with his New Look, Paris fashion leadership is regained.
1950s - 60s	Balenciaga, Givenchy, St. Laurent, André Courrèges, Pierre Cardin (France) Pucci (Italy)
1960s	Mary Quant: English designers have international influence Beginning of young designers creating for young people Mini skirts and Mod Look
1968 - 1975	Confusion in fashion direction. Paris influence declined. Ethnic influence; street fashion American jeans become international fashion
1970s	International exchange of fashion Major influence from French prêt-à-porter: St. Laurent, Kenzo (b. Japan), Rykiel, Lagerfeld (b. Germany) Italian designers important: Armani, Missoni, Krizia (Mandelli), Ferragamo shoes, Gucci handbags Halston, Calvin Klein, Mary McFadden important in American fashion
1980s	Global outlook Japanese gain international influence: shapes unrelated to body Rise of internationally known, popularly priced sportswear manufacturers such as Liz Claiborne, Esprit and Benetton Armani (Italy) sets the fashion tone for professional women Lagerfeld and Lacroix rejuvenate the French couture Ralph Lauren and Donna Karan are respected names in American fashion.
1990s	Recession has impact on high priced fashion Karl Lagerfeld is major international trendsetter with five collections Donna Karan most influential American designer

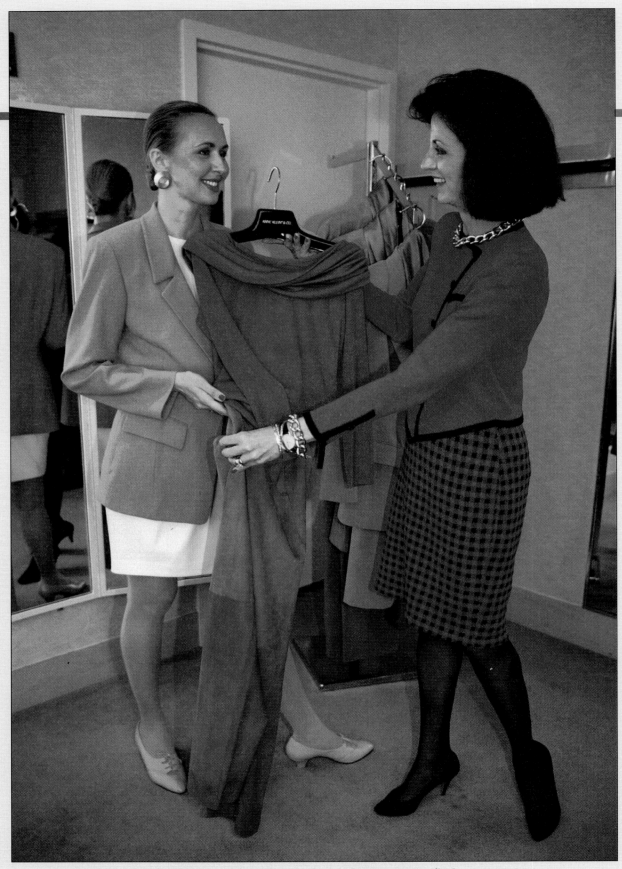

I. Magnin Personal Shopping Service director Therèse Post helps a customer find flattering and appropriate fashions. *(Courtesy of I. Magnin, photographed by the author)*

2

CONSUMER DEMAND

CAREER FOCUS

Fashion begins and ends with the consumer. Fashion executives, marketers and merchandisers including buyers, designers, and all the people involved in sales promotion continually read and learn about consumer behavior in order to get clues as to what products consumers might need or want to buy in the future. All product development and marketing decisions are based on this information. Merchandisers and marketers in touch with the needs of the consumer will be the ones most likely to succeed in an increasingly competitive marketplace.

CHAPTER OBJECTIVES

After reading this chapter you should have attained competence in the following areas:

1. Understanding how social, cultural, political, economic, and technological changes influence fashion and the fashion industry
2. Awareness of the effects of consumer demand on fashion today
3. Understanding buyer motivation.
4. Knowledge of clothing size and price ranges, style categories and clothing classifications.

*F*ashion reflects the way the consumer reacts to and is influenced by social, economic, technological, and other environmental forces in our lives. Because fashion reflects life, consumers are obviously important in determining what fashion will be and what the fashion industry will produce. Designer Bill Blass once commented, "You have to understand people to make clothes for them."[1] Retail consultant Walter Levy stresses that it is "essential to understand how consumers arrive at their decisions, to analyze the pressures that influence their priorities and the societal trends that shape what they're doing."[2] This chapter examines the predominant influences on the consumer and in turn on fashion.

Consumers are people who buy and use merchandise. *Customers*, a more specific term, are people who buy merchandise from a particular producer or retailer. As marketing strategies have become more sophisticated, manufacturers and retailers have learned to consider consumers' wants and needs. The ability to understand consumer preferences is a valuable asset at any level of the fashion industry.

MARKET SEGMENTATION

Traditionally, society has been divided by income classes. The wealthiest were the most fashionable, because only they could afford to buy expensive clothes. Eventually the clothes themselves became status symbols as people tried to prove their success by wearing expensive clothing. Today the traditional classifications have broken down. Almost all clothing is mass-produced, and almost everyone can enjoy fashion on some price level. However, *buying power*, the ability to buy (having the money) and the desire to purchase (wanting to spend the money) still has the most influence on fashion creation and fashion marketing.

Consumer goods are directed at large groups of people who have money to spend. Market research firms do *demographic studies* (statistical studies of population characteristics) for textile and apparel producers and retailers to determine what segments of the population have the most buying power. *Consumer groups* or *market segments,* which are based on age or life style (the way in which people live on the basis of location, income, education, interests, and so on), can be powerful enough to have whole new categories of clothing created for them. Although there are an infinite number of market niches, the primary demographic spending group in the United States to influence the apparel market has been the post-World War II baby boom group (people born between 1945 and 1965).

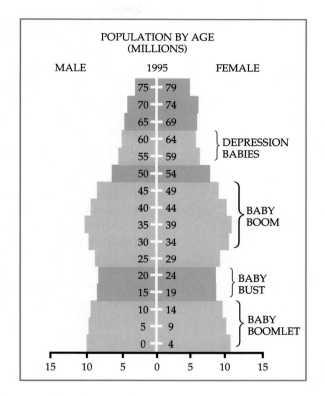

In the 1960s this youth-oriented market fostered the growth of junior sportswear, jeans, T-shirts, and other fashion fads that fit the needs of baby boomers at that time. In the 1970s, this generation became career oriented, and their new incomes and tastes were responsible for the creation of the upscale Contemporary look. By the 1980s this same demographic segment was in their thirties and forties and firmly implanted in both career and family building. Their fashion preferences changed along with their life styles and priorities. Some clever manufacturers changed their images as their customers got older. Ellen Tracy, for example, produced junior sportswear in the 1960s, contemporary in the 1970s and is now a major bridge resource.

Now the 34 to 54 age bracket is the largest segment of the population. The biggest increases will occur in the 45 to 54-age group, indicating that the 1990s is catering to a more conservative, value-and, quality oriented customer.[3] In middle age, quality gains precedence over quantity and older people tend to spend more on goods, even though they may buy less. We will "go forward into a society that is quality-driven, time-constrained, and taste-matured."[4] In the future, an aging population will contribute to slower rates of household formation and acquisition that usually belong to the young consumer. By the year 2000, half of the population will be over 40.

With the maturing of the population there is a new awareness of the consumer over 40 and 50. These age groups have previously been ignored by manufacturers, retailers, and the media. Studies show that mature people have money to spend and enjoy new products as much as anyone else. Magazines such as *Lears* and advertisements are being created for the woman and man over 40.

Whatever the age group, market researchers try to pinpoint further delineations of life styles and resulting fashion preferences. For every kind of life style there is a potential market niche. In fact, life styles and "life style dressing" have become more important than fashion trends themselves.

The major fashion market niche in the United States for over ten years has been tailored to the career woman. Since 1990, 58 percent of women over the age of 16 work outside the home. In the 25 to 44 age bracket this number exceeds 75 percent. It is of primary interest to the industry that working women spend about 35 percent more on apparel than nonworking women.[5] For these women, who divide their days between the demands of home and career, time becomes a critical issue. The busy working woman tends to favor catalog and one-stop shopping which has aided the growth of certain catalogs and super stores.

The generation born between 1965 and 1980 are often referred to as "baby busters" or "Generation X" because this age bracket is much smaller than the one preceding it.. However, these young people are fashion conscious and tend to spend more of their incomes on fashion, but at lower price levels.

Another potentially powerful consumer group is the "baby boomlet" created by the rise in birth rate that began in the early 1980s. These are the children of the postwar baby boomers. They have a computer/video orientation that will affect their buying decisions of the future. The impact of this new generation on the fashion market will be evident into the twenty-first century. The growth of the children's wear industry will continue through the 1990s and juniors will again become important before the year 2000. By 2010 the nation will be rather polarized with the postwar boomers over 50 and the new boomers under 30.

The wise merchant finds a niche and creates for or caters to that specific market. The key to fashion business success in the 1990s is market segmentation.

SOCIAL INFLUENCES THAT AFFECT CONSUMER DEMAND

Changes in people's attitudes and life styles change their fashion and buying habits. People want clothing appropriate for their interests and activities.

Only in this century have women been allowed to vote, own property, and pursue careers—and only after a long struggle. Women's new status and life styles in turn have changed the way they dress. Better-educated and exposed to new ideas, women want a wider fashion choice. To participate in sports, women need the freedom of movement provided by the clothing category of active sportswear. For jobs, women need practical and businesslike clothes such as suits. Today women's (and men's) interests point away from the fast track of the 80s towards the family and they tend to spend more money on the home than on fashion.[6] Fashion reflects the change with more femininity and softness in styling. Fashions have changed in part because women's life styles and interests have changed.

Two career families have many demands on their time, attention, and incomes.
(Courtesy of the Santimauro family, photographed by the author)

Men's clothing has changed too as a result of a refocusing of their interests and a more casual life-style. Their wardrobes—formerly limited to suits, slacks, and sport shirts—have expanded along with their activities, and increased clothing choices have made them more fashion conscious.

Men's business suits have remained basically the same; however, there has been an easing of corporate dress restrictions. Alcoa (Aluminum Co. of America), for example, found casual dressing so popular with employees that they made it an everyday option. Silicon Valley and the marketing field have led the way towards professional casualness. Many stores find that while they previously sold two suits to every sport jacket, sales have reversed. The decline in suit sales has hurt companies such as Hartmarx Corp. and Brooks Brothers Inc. Hartmarx counts on its sportswear business to keep it afloat while Brooks Brothers has introduced "suiting separates" (suit jackets and pants sold separately) which save customers the need for alterations.

CULTURAL, POLITICAL, AND ENVIRONMENTAL INFLUENCES THAT AFFECT CONSUMER DEMAND

All of the events that go on around us influence our lives directly or indirectly and affect our buying decisions.

With today's global orientation, people have become more aware of other cultures. To acknowledge this awareness, ethnic themes are again appearing in fashion collections. African influences have been seen in Christian Lacroix's collections, while Moroccan themes have been used in the collections of Gianfranco Ferre and Emporio Armani. At home, junior fashion is heavily influenced by the *Hip Hop* look which grew out of the African-American culture and rap music.

Besides international influences, we have an increasing number of immigrants that have affected the fashion market. The Immigration and Naturalization Service projects that legal immigration exceeds 700,000 per year. Stores try to cater to various tastes of ethnic groups in their area. With a multicultural society, retailers and manufacturers need to more efficiently target specific demographic and market segments.

Politics also has an affect on fashion. The Clintons are a typical baby boomer couple. While President Clinton has a casual approach to fashion, Hillary Rodham Clinton is interested in projecting a business-like appearance. However, although persued by designers, she does not have time for nor interest in fashion. Also their interest in their daughter Chelsea reflects national concerns for the family.

Concerns for the environment are also appearing in fashion collections. Designers are using natural fibers, nontoxic dyes and finishes, tencel (a rayonlike fiber made without pollution), and other nonpolluting treatments and trimmings. Fashion designer Adrienne Vittadini observes, "Fashion is a reflection of society and the concern for the earth is in the back of everyone's mind so it is only natural for designers to reveal our support through our clothes." "The use of natural fabrics and dyes along with trying to save our environment is more than a trend, it is a life style," agrees Marylou Marsh Sanders of Ecosport. "It has to be because we don't have a choice."[7]

Christian Lacroix was inspired by African motifs to design this sequined dress and fringed gold jacket.
(Courtesy of Christian Lacroix)

ECONOMIC INFLUENCES THAT AFFECT CONSUMER DEMAND

Since buying power is directly affected by income, it is important to understand all the aspects of economics and consumer spending.

Highly developed countries have the means for creating fashion demand and the facilities for fashion production. Where economic and industrial growth do not exist, as in underdeveloped countries, fashion is static or nonexistent. The economic growth of a country is reflected in the variety of its fashion.

Consumer Spending

The value-orientation of the 90s is a far cry from the conspicuous consumption of the 80s; a shift to the self-disciplined from the self-indulgent. There is a trend toward more selective buying and people are spending less on fashion. Because there is a renewed interest in the family, consumers are spending less time shopping and more time at home. Those who do buy fashion are looking for comfortable, multiple-use clothing with real value.

The amount of money consumers spend on fashion and other goods depends on their income. Income as it affects spending is measured in three ways: personal income, disposable income, and discretionary income.

Personal income is the gross amount of income from all sources, such as wages, salaries, interest, and dividends.

Disposable income is personal income minus taxes. This amount determines a person's purchasing power.

Discretionary income is income left after food, lodging, and other necessities have been paid for. This is the money available to be spent or saved at will. The increase in discretionary income enjoyed by most people in our society means that more people are able to buy fashion. Young people spend a higher portion of their income on fashion. Apparel expenses become a lower proportion of total personal expenditures for those with mortgages to pay and children to educate.

If the total personal income, disposable income, or discretionary income for an entire country is divided by its total population, the result is the average *per capita income* in each category.

Income is related to the economic situation. Although incomes in the Western world have risen in recent years, so have prices. Thus, income is meaningful only in relation to the amount of goods and services it can buy—its purchasing power. Inflation, recession, the international value of currency, and productivity affect purchasing power.

Inflation

In an inflationary period such as the United States experienced in the 1980s, people earn more money each year but higher prices and higher taxes result in little or no real increase in purchasing power.

The fashion emphasis of the 80s was the power suit and luxury which reflected society's obsession with money and self-interest.

Recession

A recession, such as we experienced in the early 1990s, is a cycle beginning with a decrease in spending. Many companies are forced to cut back production, which results in unemployment and a drop in the gross national product (GNP). Unemployment furthers the cycle of reduced spending.

When the economic situation is unstable, the fashion picture is also unstable. Not only is money in short supply, but people seem to be confused about what they really want. In an economic upswing colors are basically cheerful and happy, but they have a grayed palette when the economy is in trouble.[8] In a recession, people are likely to buy conservatively or at least buy fashions they believe to be of lasting value. The success of discounters such as Wal-Mart during the last recession demonstrated consumers' quest for value.

The International Money Market

When the dollar is strong against other currencies, American consumers are able to buy international merchandise more cheaply. At the same time, American industry is hurt because foreign merchandise competes with domestic goods. Moreover, American exports become too expensive for other countries to buy.

On the other hand, when the dollar is weak (loses value relative to other currencies), then foreign countries can buy American goods more cheaply. This situation encourages American fashion manufacturers to export and is good for business. However, imported goods are more expensive for us to buy.

Labor Costs

As people receive higher salaries and live better, the cost of making garments increases. Rising labor costs have made clothing very expensive and caused many manufacturers to search for cheaper sources of labor in the Far East, the Carribean Basin, and elsewhere. This has caused a controversy between American workers, who feel that their jobs are in jeopardy, and proponents of free trade, who feel that consumers should pay the lowest possible price for quality merchandise (see Chapters 10 and 12). Increased labor costs may also eventually dictate more simply constructed clothes, barring additional technical breakthroughs that reduce the impact of such costs.

TECHNOLOGICAL INFLUENCES AFFECTING CONSUMER DEMAND

The technological advances that have made mass production of fashion possible are accelerating faster every decade.

Mass Production

Modern technology makes production more efficient, resulting in reduced costs and cheaper prices. Obviously, mass production would have been impossible without early technological inventions such as the sewing

A computer-automated knitting system (*Courtesy of Sauer Textile Systems*)

machine. The development of modern production machinery such as power sewing machines and cutting tools has streamlined the process of manufacturing further. Today's power machines can run faster than a car engine, sewing over 5000 stitches per minute. Modern cutting techniques include the use of computers, water jets, and laser beams. Now, computer technology has again revolutionized manufacturing with computer-aided design, pattern making, grading, cutting, unit and modular production systems, pressing and distribution systems (see Chapter 10).

Technological inventions in the textile industry actually triggered the Industrial Revolution. Since then, advances in spinning, weaving, and knitting processes have fostered the growth of our giant textile industry. Today, computer-aided spinning, design, weaving, knitting, dyeing, and finishing allows our domestic industry to compete with textile imports. Modern agricultural developments have improved the quantity and quality of natural fibers. Technological research has made man-made fibers possible, as well as finishes that change fabric characteristics. Certain companies have developed environmentally friendly finishing detergents, cleaners, oils, and dyes for the textile industry (see Chapter 5).

Transportation

The speed and convenience of airplane and automobile travel have given Americans a mobile life style. Travel necessitates a lightweight, seasonless, packable wardrobe. Textile technology answered the need with easy-care, wrinkle-free fabrics.

Modern transportation has also made the logistics of the fashion business more manageable. Improved trucking, railroad, and air-freight services quickly bring the newest fashions to the retail store and the consumer. Air freight has also made imports possible as it allows for quick shipments of samples during product development.

Communication

Modern communications media bring different cultures into contact, making people more aware of other life styles and modes of dress. In the past it took many months for a magazine showing the latest Paris fashions to be printed and shipped to its readers. Now, television brings fashion from around the world into our homes almost instantly. A facsimile (fax) machine can send a fashion sketch around the world in 40 seconds. As a result, the public is quickly made aware of the existence of new styles. Thus, one of the greatest impacts of modern communication on fashion is the acceleration of change. Communication also speeds up the process of copying, so that merchandise produced throughout the world tends to look similar. The sooner a fashion saturates society, the sooner people tire of it and want change.

Computers

Computers have become indispensable to the fashion industry. If fed the proper information, they bring order out of chaos by keeping track of or controlling planning, production, inventory, and distribution. In the planning stages, merchandisers rely on historical data immediately available to them by computer. In textile production computers run machines and aid in textile design. In apparel manufacturing they make, grade, and lay out patterns in marker form. Some manufacturers are experimenting with computers for apparel design. At all levels of the industry, computers control inventory and distribution, making it easier and faster to fill orders or replenish stock. Domestic textile and apparel producers and retailers use *electronic data interchange* (EDI), a cooperative computer linkage between all levels of the industry to try to speed up the response to consumer demand (see Chapters 6 and 12 for detailed explanations). Using computers whenever possible saves time on paperwork and leaves designers, merchandisers, sales reps, and buyers free for creative thought.

MARKETING INFLUENCES AFFECTING CONSUMER DEMAND

Marketing influences consumer demand but manufacturers and retailers have learned that consumer demand must affect marketing.

The retailing visionaries of the late nineteenth century such as Montgomery Ward, who guaranteed "satisfaction or your money back," and Marshall Field, who advised his employees that "the customer is always right," began the development of consumerism in the United States. *Consumerism* refers to the right of the buying public to be protected against unfair marketing practices. The economic depression of the 1930s stimulated reform legislation in many areas of business, including consumer protection.

After the artificial economy created by World War II and the enormous growth of advertising in the 1950s, the consumer-rights mood of the 1960s spurred additional government protection for the consumer. Today there are many regulations, especially concerning truthful and factual labeling. Since World War II there has been a major change in marketing philosophy. Because more people have more money to spend, competition for consumer dollars has become very intense. Analysts have developed sophisticated marketing research methods to determine consumer wants and needs; managers have emphasized product development to answer those needs; and fashion firms are spending vast amounts of money on increased advertising and other promotional activities to create consumer demand.

Advertising and publicity stimulate the public's desire for new clothes. The ultimate achievement of advertising is to establish the identity of a particular brand name or store so solidly that it will be preferred over the competition. However, there is a limit to which sales promotion can win acceptance for a fashion. If the public is not ready for a product, or is tired of it, no amount of advertising or publicity can gain or hold its acceptance. Marketing in the future must be more concerned with finding out what the consumer really wants.

MOTIVES FOR CONSUMER BUYING

To help them make styling and merchandising decisions, designers, buyers, and other industry executives try to understand consumer motivation.

In the past, most people bought new clothes only when a need arose, for a very special occasion, or because their old clothes wore out. The average person simply could not afford to buy more than the basic necessities. In Western society today, discretionary income is larger, and people can buy new clothes rather frequently. Therefore, buying motives have changed; we are able to buy clothes because we want or like them. Since women control 80 percent of all spending on apparel, understanding their motivations is the biggest challenge facing the industry. [9] Buying motives vary from consumer to consumer and from day to day. They include the desire to

Be fashionable.
We may buy new clothing to feel that we are trendy or at least in the mainstream of fashion. We discard clothing that is still wearable only because it is out of fashion.

Be attractive.
We want clothes that are flattering, that make us look our best, or show off our physical attributes.

Impress others.

We may want to project a successful image or establish our identities with fashion. We may want to exhibit our taste level or income level through clothing. Expensive brands have even served as status symbols.

Be accepted by friends, peer groups, or colleagues.

Average Americans have conservative tastes; we do not want to differ from our peers. We may want to project a life style that others will identify with. Buying patterns suggest that we like some direction or guidance as a framework for our choices.

Fill an emotional need.

New clothes often help us feel better psychologically. Being secure in the feeling that we are wearing appropriate fashion helps us feel confident and self-assured. This motive, however, may often lead to impulse buying (buying without careful consideration). Corrin Corbin, director of apparel marketing for DuPont, points out, "the apparel chain must work together as a team to fully understand and satisfy the consumer on her deepest emotional levels if the industry is to prosper."[10]

CRITERIA CONSUMERS USE IN FASHION SELECTION

To determine the acceptability of fashion, both designers and manufacturers find it helpful to consider the criteria used for its selection.

Elements of fashion appeal draw the consumer's attention to a fashion. There are also practical considerations, including quality and price, that the consumer usually evaluates before making a purchase.

Esthetic Considerations

Esthetic considerations are very similar to the elements of design, but they are from the purchaser's point of view rather than the creator's:

Appearance

The overall appearance of a garment or accessory is what first attracts a potential buyer. This can depend on hanger or display appeal as well as the following:

Color

People relate very personally to color, usually selecting or rejecting a fashion because the color does or does not appeal to them or flatter their own coloring.

Texture

The surface interest in the fabric of a garment or accessory is called texture. Texture usually gives a clue as to fiber content.

Style

The elements that define a style include line, silhouette, and details. A consumer's selection is frequently influenced by his or her opinion of what is currently fashionable.

Practical Considerations

Price

Value has become the key to merchandising for the 90s. Consumers want the best product at the best price. Price is probably the most important practical consideration for the average consumer. The consumer must compare the total worth of the esthetic aspects of the garment or accessory with the retail price and his or her own budget.

Fit

The try-on is a crucial step in the consumer's selection of a garment because sizing is not a guarantee of fit. The U.S. Department of Commerce has tried to set sizing standards, but each company tends to vary somewhat. Each company tries its sample garments on models who are typical of the company's customers. However, it is difficult to set size ranges and grading rules to fit every figure. The fitting room try-on further enables the customer to judge if esthetic elements enhance his or her figure type or general appearance.

Comfort

Traditionally we have always needed clothes to keep us warm in cold weather or cool in warm weather. The trend for physical fitness and the increase in travel have made us also aware that we want clothes that are comfortable not only for sports and exercise but also to move in, sit in, travel in, and so on.

Appropriateness

It is very important that a particular garment be suitable or acceptable for a specific occasion or for the needs of the consumer's life style. Appropriateness includes considerations of job

Checking the fit of a suit at Gieves & Hawkes, London. *(Courtesy of Gieves & Hawkes)*

versus leisure, day versus evening, type of event, and the age of the wearer. Impulse shoppers do not consider appropriateness and therefore purchase many items that do not fit into their wardrobes.

Brand or designer label

Brands are a manufacturer's means of product identification. Some consumers buy on the basis of a particular brand's reputation, often as a result of heavy advertising. Designer Giorgio Armani states, " I believe consumers in some markets are becoming less brand-conscious. A brand name is important as long as it is combined with a proper relationship of quality and price."[11]

Fabric performance and care

The durability of a garment or accessory and the ease or difficulty of caring for it are often factors in selection. Many consumers prefer easy-care, wash-and-wear fabrics, although designer and contemporary customers may not mind paying for dry cleaning the natural fiber fabrics they prefer. Easy care and durability are of special concern in children's wear and work clothes. Government regulations now require fiber content and care instruction labels to be sewn into apparel.

Quality

Consumers are demanding higher quality of construction, stitching, and finishing. Quality standards fall as labor costs rise and managements favor more profitable balance sheets. Unfortunately, many consumers cannot and do not bother to evaluate workmanship. The junior customer cares little about quality; she is likely to throw away a garment before it wears out. The designer or missy customer, on the other hand, generally considers clothing an investment and may not mind spending more for the lasting qualities of fine detailing and workmanship. Some consumers may look for a particular brand or name on the basis of a reputation for quality. Today consumers are demanding higher quality to give them value for the prices they are expected to pay.

Convenience

With time and energy in increasingly short supply, consumers are looking for ways to make shopping easier. Consumers are spending less time shopping than before. They want to find what they want and find it easily and quickly. They want service when they need it.

MEETING CONSUMER DEMAND

To meet consumer demand and changes in consumer life styles, manufacturers and retailers have developed various size and price ranges as well as categories for styling and clothing type.

Size Ranges

Each size range caters to a different figure type. However, every manufacturer has its own interpretation of sizing. At least today's sizing is generally based on more realistic measurements than the idealistic sizing of the past. Designer clothes are generally cut for a slim figure while more traditional companies may have roomier cuts. On the other hand, designer collections are often deliberately cut one size bigger so that the

TABLE 2-1
Typical Relationship of Style, Size, and Price Ranges in Women's Wear

style range	styling	age	size range	figure	price range
Designer	Unique, top-name designer fashion	25 and up	Missy 4-12	Missy: fully developed but slim	Designer
Bridge	Designer fashion	25 and up	Missy 4-12	Missy: slim, developed	Better and up
Missy	More conservative adaptations of last season's designer looks	25 and up	Missy 4-14	Missy: developed, 5'7" block	Better to budget
Petites	Same as missy	25 and up	Petite 0-14	Missy: under 5'4"	Better to budget
Women's or large sizes	Same as missy plus some junior looks	18 and up	16-26W or 16-26WP	Missy: large size + some petite	Better to budget
Contemporary	Trendy	20-40	Missy 4-12	Missy: slim	Better to better
Junior	Youthful, trendy, figure-conscious	15-25	Juniors 3-15	Not fully developed	Better to budget

customer who spends a lot of money can feel good about being a size 10 instead of a 12. Some mainstream manufacturers such as Liz Claiborne try to accommodate a variety of body types by using a variety of fit models.

Junior customers, sizes 3 to 15, have a less developed figure and a shorter back-waist length (a higher waistline) than missy figures. This figure is usually equated with a young customer.

Missy sizes 6 to 16 (or 4 to 14) are for the mature female figure, usually based on a 5'7" height. In missy separates, some blouses and sweaters are sized 30 to 36 (8 to 14), or small, medium, and large. Manufacturers have adopted more realistic average measurements to replace the former idealistic standards.

Petite sizes are created for the woman under 5'4". Sizes range from 0 to 16. Most manufacturers limit the size range to 4 to 14 or 6 to 16.

Large or women's sizes range from 14W to 32W but are usually limited to 16W to 26W. Sometimes these sizes are represented as 1x (16 to 18), 2x (20 to 22) and 3x (24 to 26). Large-size petites are marked WP.

Fashion for petite and large size women was virtually ignored by the fashion industry until 1977. Statistics now show that 54 percent of the total female American population wears either petite or large sizes, or both. Among women over 40, 40% wear size 14 or larger.[12] Prestigious manufacturers such as Liz Claiborne (Elizabeth) and Givenchy (En Plus) are now catering to special sizes.

It is difficult to compare international sizing. Particularly in France, the sizing is not always standard. Table 2–2 shows how confusing a comparison of sizes between countries can be.

TABLE 2-2
International Dress Sizes

American	8	10	12	14	16
British	10	12	14	16	18
French	40	42	44	46	48
German	36	38	40	42	44

Mother and child coordinating fashions from Laura Ashley. *(Courtesy of the Laura Ashley Company)*

Men's suits range in size from 36 to 44 (with additional large sizes to 50), based on chest measurements. Lengths are designated after the size number: R for regular, S for short, and L for long. European sizes are 46 to 54 (add 10 to each American size). Young men's sizes, equivalent to junior sizes for women, have a narrower fit in the jacket and hip and a shorter rise in the trouser than regular men's sizes.

Dress shirts are sized by collar measurement (inches in America and centimeters in Europe) and sleeve length. Sport shirts are sized in small, medium, and large. Trousers are sized by waist and inseam measurements.

Children's wear is sized by age group. Infant sizes are based on age in months, usually 3, 6, 9, 12, and 18. However, since development varies so much from child to child, many manufacturers are now also identifying weight ranges on their labels. In Europe, sizes are based on the length of the baby or the height of the child.

Toddler clothes, for the child who has learned to walk, are sized 1 to 3; children's sizes are 3 to 6X. At this point, sizes separate for boys and girls. Girls' wear comes in sizes 7 to 14; the developing adolescent girl wears subteen 6 to 14; and the young teen wears young junior 3 to 13. Boys' sizes are patterned after men's and include sizes 7 to 14 and 14 to 20.

Price Ranges

A garment should give good value for its price. There are many price ranges, each with a different level of customer expectation. As the price goes up, the customer expects higher quality in fashion, fabric, fit, and finish.

Designer garments became so expensive that a new category was created to "bridge" the gap between designer and better missy or contemporary. Bridge lines include less expensive designer lines such as Anne Klein II, DKNY (Donna Karan New York), Ungaro Ter, and collections aimed specifically at this price range such as Ellen Tracy (Linda Allard, designer), Adrienne Vittadini, and Laurel (Escada, Germany).

Each garment manufacturer generally specializes in one price range. The designer and merchandiser must consider the cost of every fabric trim or construction detail that goes into a garment. Costs must fit into a specific price range. In turn, each retail store has various departments, from budget to designer, again classified by price range.

Style Categories

Both women's dresses and women's sportswear currently come in style ranges as well as size ranges. Some of the terms overlap because style ranges grew out of size ranges.

Women's wear

Couture. This term is reserved for fashion that is made to order to fit an individual client's measurements. These clothes are often the most innovative, and definitely the most luxurious and the most expensive. These collections are supported by ready-to-wear lines and licensing (see Chapter 8). Examples of couturiers include Karl Lagerfeld, Christian Lacroix, Yves St. Laurent, Valentino, and Arnold Scassi.

Designer. Name designers who design for their own labels each have their own distinctive style. Success and high prices allow them to keep using the best of fabrics and workmanship. Today designer ready-to-wear is as expensive as couture used to be. Examples of name designers include Giorgio Armani, Donna Karan, Ralph Lauren, and all the designers listed in Chapter 8.

Bridge. These fashions are simply a step down in price achieved by using a less expensive fabric or different production methods. Some designers have second lines such as Donna Karan's DKNY, Versus from Versace, or Emporio from Armani. Other collections such as Nicole Miller, Dana Buchman or Adrienne Vittadini are designed specifically for the bridge market.

Missy. These are more conservative adaptations of proven or accepted designer looks. They utilize less expensive fabrics, less extreme silhouettes, and use mass production methods. There is a wide range of prices, however, between better and budget missy styling. Vendors include J H Collectibles, Liz Claiborne dresses and sportswear, and Leslie Fay.

Contemporary. This is a fashion forward category aimed at the young woman, with a limit on price but more sophisticated than junior. Resources include Sue Wong and Barbara Barbara dresses or Max Studio (Leon Max) sportswear.

Designers Karen Harman and Dana Buchman check fit on a model at Dana Buchman, a bridge resource. *(Courtesy of Dana Buchman)*

Junior. Young styling is heavily influenced by the rock music scene and by street fashion in Europe. It also tends to be body conscious. Labels include Betsey Johnson dresses, and Guess, Cherokee, and Esprit sportswear .

Within retail stores, there is sometimes overlapping between these categories. Classification depends in part on the size, location, and clientele of the retail store. Liz Claiborne might be found in the missy department of a large downtown specialty store and in the better department of a suburban branch store. Also, as manufacturers or designer names grow in popularity, they add lines in other categories and price ranges. Claiborne, Vittadini, and Tracy now make dress lines in addition to sportswear. Claiborne also added men's wear collections. Ralph Lauren, on the other hand, started by designing men's wear and later added women's. Other designers and manufacturers such as Jessica McClintock and Generra have added children's wear lines.

Men's wear

Designer and contemporary styling has carried over to men's wear, especially since many designers are doing both. Men have become more involved, interested, and discerning when purchasing their wardrobes. In the suit category, couture and designer suits are conservative in styling but express elegance with superior quality fabrics, construction and cut. The finest suits are manufactured by Oxxford in America, Zegna in Italy and the London Savile Row and Hong Kong tailors.

Whereas suits represented 50 percent of men's wardrobe purchases in the past, now men's clothing purchases are divided equally among suits, furnishings and accessories (shirts, ties, pajamas, and so on), and sportswear (leisurewear and activewear). The styling creativity and diversity now apparent in men's sportswear is a result of an easing of company dress codes as well as a desire on the part of men to have appropriate clothing for their leisure activities. Although MacGregor and Pendleton did some men's coordinates as early as 1947, it was Ralph Lauren's Polo line that made the look really catch on. Today popular brands include Lauren, Tommy Hilfiger, Hugo Boss and the popular priced Levi Dockers.

Casual suit from Giorgio Armani.
(Courtesy of Giorgio Armani)

Children's wear

In the case of small children, the consumer is a parent, grandparent, or other adult. Children's clothes, too, can be designer, trendy, traditional, or simply functional. Older children, however, have more definite opinions on what they want to wear, partly because of advertising, television exposure, and peer group pressure. This development has had an effect on styling in that some childrens' wear manufacturers follow junior trends.

Clothing Classifications

There have never been as many types of clothing as we see today. Variety in dress has resulted from changes in our habits and roles and the increase in our purchasing power. We now have clothing for all occasions and life styles. Retail stores have separate departments for each category of clothing. There are also many subclassifications based on price range and the target customer's life style.

Women's wear

Women's clothes have many classifications including lingerie, dresses, evening clothes, suits, outerwear, and sportswear. There are also specialty categories, such as bridal gowns and maternity clothes. In addition, there is a huge array of accessories such as scarves, hats, handbags, footwear, and hosiery.

Lingerie includes undergarments, bodywear, sleepwear, and loungewear.

Dresses range from the very tailored with crisp lines for wearing on the job, to the very softest with gathers and ruffles for dressy occasions.

Evening clothes vary from party pajamas to long and short cocktail dresses to opulent gowns.

Suits are jackets and skirts sold together as units. Suits range from the soft "dressmaker" suit to the strictly tailored.

Outerwear includes coats, capes, and jackets with a primarily protective function.

Sportswear is actually any combination of tops and bottoms that are priced separately so that the customer can combine them as desired. Sportswear lines are organized as separates or as coordinated sportswear, pieces intended to be mixed and matched. As leisure time and discretionary income have increased, the sportswear category has grown tremendously. American designers have excelled in this category to suit informal American life styles.

Active wear from Fila's Pro Beach Volleyball Collection. *(Courtesy of Fila Sports Inc.)*

Active sportswear grew from virtually nothing to a very important category as consumers learned to value physical fitness. Active sportswear is designed for movement and worn for a variety of fitness activities or sports. Tops and bottoms can be priced separately or as a unit.

Men's wear

There are now almost as many categories available to men as to women. Many designers and manufacturers, including Ralph Lauren and Hugo Boss, have lines in several men's wear categories. Stores have increased square footage in men's wear areas and use aggressive display and promotional techniques to lure their increasingly fashion-wise male customers.

Tailored clothing includes suits, overcoats, topcoats, sport coats, and separate trousers for business, leisure, and evening wear.

Furnishings are shirts, neckwear, sweaters, tops, underwear, socks, robes, and pajamas.

Sportswear comprises knit or woven related separates that fill the demand for more leisure and casual wear.

Active sportswear includes all garments needed for sports or exercise such as windbreakers, ski jackets, jogging suits, and tennis shorts.

Work clothes such as overalls, work shirts, and pants required by laborers have become an option for leisure wear.

SUMMARY

Outside influences such as changing life styles, general economic conditions, labor costs, government legislation, technological advances, and advertising strongly influence consumer demand. If we learn to observe these influences, we can better foresee the need for new fashion development and change.

The consumer is an important determinant of what fashion is and what the industry produces. The fashion industry caters to powerful consumer groups because they have the most discretionary income. Consumers buy clothes for many reasons, including the desire to be fashionable, attractive, impressive, accepted, or emotionally fulfilled. Considerations in fashion selection include color, texture, style, price, fit, appropriateness, brand name, fabric performance, and workmanship. Clothing is available in various size ranges, price ranges, style ranges, and classifications.

CHAPTER REVIEW

Terms and Concepts

Briefly identify and discuss the following terms and concepts:

1. Consumer demand
2. Buying power
3. Life styles
4. Demographics
5. Market segmentation
6. Disposable income
7. Discretionary income
8. Inflation
9. Recession
10. International money market
11. Labor costs
12. Buyer motivation

Questions for Review

1. How have life style changes affected dress for both men and women?
2. How have demographics influenced fashion?
3. Discuss people's motives for buying clothes.
4. How have technological advances changed fashion?
5. What does a consumer consider when buying clothes?
6. What are the similarities and differences between style ranges and price ranges?

Projects for Additional Learning

1. Analyze fashion styles today. Is there one that you feel is the result of a social, economic, or technological influence? Briefly explain in writing the reasons for your choice.
2. Interview two executive businesswomen and find out what types of clothes are in their wardrobes. How do their needs affect the fashion industry?
3. Market research: In a local specialty or department store, compare two dresses of similar price from different manufacturers. Try them on. Which has the better fit? Compare fabric choices in terms of quality and suitability to the design. Compare styling. Which is more innovative? Taking all factors into consideration, which dress is the better buy? Discuss your findings in a written report. Describe or sketch the two garments.
4. In a local specialty or department store, compare the missy and junior departments. List the major manufacturers in each department, and explain the general styling differences.

NOTES

[1] Quoted in Barbra Walz, *The Fashion Makers* (New York: Random House, 1978), p. 47.

[2] Walter K. Levy, chairman, Walter K. Levy Associates, Inc., speaker, Fashion Group Forum, January 19, 1989.

[3] Joseph H. Ellis, partner, Goldman Sachs & Co, speaker, Fashion Group Forum, January 19, 1989.

[4] John J. Shea, president, CEO, Spiegel, speaker, Fashion Group Forum, January 19, 1989.

[5] NPD Special Industry Services, "Career Smart," DuPont brochure, p. 3.

[6] Ellen Sideri, the Fashion Works, interview, February 3, 1993.

[7] As quoted by Sara Fiedelholtz, "Makers: Boosting Figures the Natural Way," *Women's Wear Daily*, Sportswear Trends Supplement, February 1993, p. 2.

[8] Margaret Walch, associate director, Color Association of the United States, interview, February 3, 1993.

[9] Marvin Klapper, "DuPont Tells Industry: Learn What Women Want," *Women's Wear Daily*, October 16, 1991, page 2.

[10] Ibid.

[11] As quoted in "Europe Design Houses Face a New World," *Women's Wear Daily*, June 15, 1992, p. 6.

[12] Amy Feldman, "Hello Oprah, Good-bye Iman," *Forbes*, March 16, 1992, p. 116.

Gianfranco Ferré with models at the showing of his new Dior collection.
(Courtesy of the house of Dior)

3

FASHION CHANGE AND CONSUMER ACCEPTANCE

CAREER FOCUS

Because consumers are the end users of fashion, marketers, merchandisers, and designers must consider them in the planning stages in order to make fashions they will want. Consumers' wants and needs create a cycle of consumer demand, industry catering to that demand, and finally consumer acceptance of merchandise offered in the market.

CHAPTER OBJECTIVES

After reading this chapter you should have attained competence in the following areas:
1. Understanding the dimensions of fashion
2. Identifying the phases and lengths of fashion cycles and relating them to consumer acceptance
3. Comprehending fashion adoption theories in relation to consumer acceptance

*T*he first part of this chapter discusses fashion acceptance and rejection, a cycle that creates fashion change. The remainder of the chapter covers the consumer's connection to these cycles and how this relates to forms of fashion adoption. First, we will discuss terms associated with the acceptance of fashion.

FASHION TERMS

All fashion executives use the following terms daily to discuss the aspects of fashion.

Fashion is the style or styles most popular at a given time. The term implies three components: style, acceptance, and timeliness.

Style is any particular characteristic or look in apparel or accessories. Designers interpret fashion ideas in this creative form and offer them to the public. Style may come and go in fashion, but a specific style always remains the same, whether it is in fashion or not. For example, the shirtwaist style will not always be in fashion, yet it will always involve the same styling and details, which make it a shirtwaist. Individuals can also develop their own personal style of dress.

Acceptance implies that consumers must buy and wear a style to make it a fashion. World-renowned designer Karl Lagerfeld remarked, "There's no fashion if nobody buys it."[1] Acceptance by a large number of people makes a fashion important. However, different groups adopt different fashions. What appeals to a junior customer would probably not appeal to a missy customer. Designers plan styles to appeal to certain consumer groups—their particular customers. It is then up to the public to decide whether these styles will become fashion.

Timeliness indicates change. What is in fashion one year will be out the next. Karl Lagerfeld said, "What I like about fashion is change. Change means also that what we do today might be worthless tomorrow, but we have to accept that because we are in fashion. There's nothing safe forever in fashion...fashion is a train that waits for nobody. Get on it, or it's gone."[2] Change is what makes the fashion business exciting. However as designer Liz Claiborne commented, "I don't think fashion is as exciting as it once was. Women aren't waiting breathlessly to see what's going to be in for fall or spring."[3]

Many people criticize the fickleness of fashion, saying that fashion changes only to stimulate buying. And it is true that if fashion never changed, the public would not buy clothing so frequently. However, fashion is one way for us to visually express our relationship to current events and to life itself. The speed of change is influenced by modern communication, marketing, advances in mass production, greater discretionary income, and the seasons. Because fashion is a product of change, a sense of timing (the ability to understand the speed of acceptance and change) is an important asset for anyone involved with product development or marketing in the fashion industry. Italian designer Valentino remarked, "A great designer should always try to create trends with an eye on the market; that's why often timing is the key to a successful idea."[4]

FASHION EVOLUTION

Fashion doesn't change overnight. It is important for designers and merchandisers to understand fashion cycles because they explain the acceptance of fashion and are directly related to buying and selling cycles.

Generally, fashion changes evolve gradually, giving consumers time to become accustomed to new combinations and looks. Designers arrive at new fashion looks by changing design elements such as line, shape, color, fabric, and details, and their relationship to one another. For instance, a change in skirt length might in turn affect the proportion of the skirt to the bodice of a dress. In men's suits, the narrowing of lapels tends to reduce the scale of other details and accessories, such as neckties (see Chapter 9 for a discussion of design).

Fashion Cycles

Consumers are exposed each season to a multitude of new styles created by designers. Some are rejected immediately, often by the press or by the buyer on the retail level, but others are accepted for a time, as demonstrated by consumers purchasing and wearing them.

The way in which fashion changes is usually described as a fashion cycle. It is difficult to categorize or theorize about fashion without

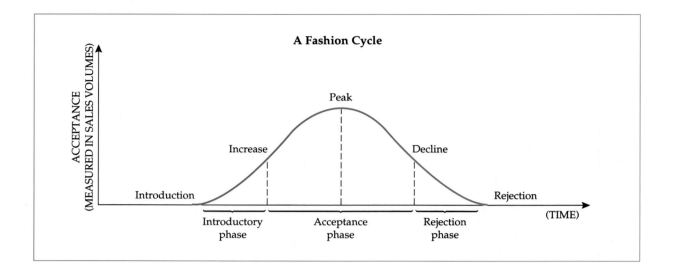

A Fashion Cycle

oversimplifying. Even so, the fashion cycle is usually depicted as a bell-shaped curve encompassing five stages: introduction, rise in popularity, peak of popularity, decline in popularity, and rejection. The cycle can reflect the acceptance of a single style from one designer, or of a general style such as the miniskirt.

Introduction of a style

Designers interpret their research and creative ideas into apparel or accessories, and then offer the new styles to the public. New creations referred to as the "latest fashions" may not yet be accepted by anyone. At this first stage of the cycle, fashion implies only style and newness.

Most new styles are introduced at a high price level. Designers, whose names are respected for both their creativity and their sense of timing, are often given financial backing and allowed to design with very few limitations on creativity, quality of raw materials, or amount of fine workmanship. Naturally, production costs are high and only a few people can afford the resulting garments. Designers such as Arnold Scaasi, Karl Lagerfeld, or Valentino do not expect to sell their couture collections in quantity. Production in small quantities gives a designer more freedom, flexibility, and room for creativity.

Increase in popularity

As the new fashion is purchased, worn, and seen by more people, it may begin to rise in popularity. In the case of an expensive item, sales will never be high, but the item may be the most popular in a designer's collection or even the most popular of all high-priced new fashions. Most couture and high-priced designers now have secondary, bridge and/or diffusion lines at lower prices so that they are able to sell their designs in higher quantities.

The popularity of the style may further increase through copying and adaptation. Some designers or stylists may modify a popular style to suit the needs and price range of their own customers. Some manufacturers may try to copy it with less expensive fabric and less detail it order to sell the style at lower prices.

Peak of popularity

When a fashion is at the height of its popularity, it may be in such demand that many manufacturers copy it or produce adaptations of it at many price levels. Styles with great appeal are produced in many variations. Donna Karan's "cold shoulder" long dress (a black dress with bare shoulders) was worn by Liza Minelli at the 1992 Academy Awards which was seen around the world. First Lady Hillary Rodham Clinton wore the same style in January 1993 and also received media exposure. By spring of 1993, the look had been copied in many forms of dresses and tops at all price levels and could be found in many stores and catalogs.

Volume production requires a likelihood of mass acceptance. Therefore, most volume manufacturers follow sales trends (see Chapter 4) because their customers want clothes that are in the mainstream of fashion.

Decline in popularity

Eventually, so many copies are mass-produced that fashion-conscious people tire of the style and begin to look for something new. Consumers still wear garments in the style, but they are no longer willing to buy them at regular prices. Retail stores put such declining styles on sale racks, hoping to make room for new merchandise. Of course, most of the merchandise on the sale racks was never acceptable.

Rejection of a style, or obsolescence

In the last phase of the fashion cycle, consumers have already turned to new looks, thus beginning a new cycle for another style. The rejection or discarding of a style just because it is out of fashion is called *consumer obsolescence*.

Length of Cycles

Although all fashions follow the same cyclical pattern, there is no measurable timetable for a fashion cycle. Some fashions take a short time to peak in popularity, others take longer; some decline slowly, others swiftly. Some last a single selling season, others last several seasons. Certain fashions fade quickly, others never completely disappear.

Classics

Some styles never become completely obsolete, but instead remain more or less accepted for an extended period. A classic is characterized by simplicity of design, which

A classic suit from Gieves & Hawkes, Savile Row, London. *(Courtesy of Gieves & Hawkes)*

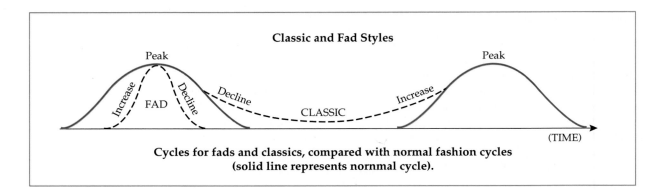

Classic and Fad Styles

Cycles for fads and classics, compared with normal fashion cycles
(solid line represents nornmal cycle).

keeps it from being easily dated. An example is the Chanel suit, which peaked in fashion in the late 1950s and enjoyed moderate popularity again in the 1980s. In the interim, the house of Chanel in Paris, as well as other manufacturers such as Adolfo, have produced variations of these suits for a small, dedicated clientele.

Fads

Short-lived fashions, or fads, can come and go in a single season. They lack the design strength to hold consumer attention for very long. Fads usually affect only a narrow consumer group, begin in lower price ranges, are relatively simple and inexpensive to copy, and therefore flood the market in a very short time. Because of the market saturation, the public tires of them quickly and they die out. A fad of the early 1980s was the *Punk* look from England, which heavily influenced the international junior market .

Cycles within cycles

Design elements (such as color, texture, silhouette, or detail) may change even though the style itself remains popular. Jeans became a fashion item in the late 1960s and remained so. Therefore, their fashion cycle was very long. However, various jean silhouettes—including bell, cigarette, and baggy —came and went during that time.

Interrupted cycles

Consumer buying is often halted prematurely because manufacturers and retailers no longer wish to risk producing or stocking merchandise that will soon decline in popularity. This is obvious to consumers who try to buy summer clothes in August.

   Sometimes the normal progress of a fashion cycle is interrupted or prolonged by social upheaval, economic depression, or war. Consider the large-shouldered, wedge-shaped silhouette in women's fashion which began in the 1930s. Because people were concerned with things more important than fashion during World War II, the same silhouette continued, without the normally expected decline, for the duration of the war. The *New Look* of 1947, with its sloping shoulders, tiny waists, and longer skirts, was a radical change because the old cycle had been unnaturally prolonged.

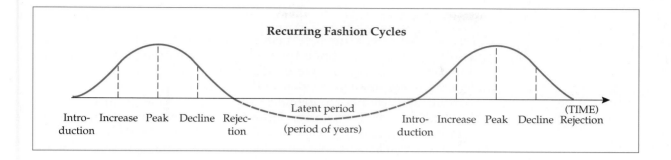

Recurring Fashion Cycles

Recurring cycles

After a fashion dies, it may resurface. Designers often borrow ideas from the past. When a style reappears years later, it is reinterpreted for a new time; a silhouette or proportion may recur but it is interpreted with a change in fabric and detail. Nothing is ever exactly the same—yet nothing is totally new. In the 1980s the padded-shoulder silhouette of the 1940s was reintroduced. However, the use of different fabrics, colors, and details made the look unique to the 1980s.

CONSUMER IDENTIFICATION WITH FASHION CYCLES

How the customer relates to the phases of fashion cycles has to do with the consumer group they belong to and gives clues to fashion needs.

Taste

An individual's preference for one style or another is referred to as taste. Good taste in fashion implies sensitivity to what is beautiful and appropriate.

Acceptance by the public does not prove that a design is necessarily beautiful, only that its timing is right. Really beautiful design may not be accepted by the general public at all, because it is too expensive. Watered-down copies often lose their original beauty (which is not to say that all couture is beautiful).

Consumer Groups

Consumers can be identified with various stages of the fashion cycle. Fashion leaders buy and wear new styles at the beginning of their cycles; others tend to imitate. Because of differences in taste, what is fashionable for one group is not for another group. What is in fashion for trendsetters is too extreme for most of us who are followers. Manufacturers and retailers may also be identified as fashion leaders or followers, depending on which consumer groups they target.

Fashion leaders

The people who look for new fashion and wear it before it becomes generally acceptable are often referred to as fashion leaders. They may give impetus to a certain style by discovering and wearing it. Fashion leaders are a very small percentage of the public.

Most fashion leaders are members of higher income groups, because quality high fashion is expensive. However, there are many young people who dare to express their own style although they do not get international attention. Fashion leaders constantly look for interesting new styles, colors, fabrics, and ways to accessorize their clothes. They try to find unique

Many young women want to look like celebrity model Claudia Schiffer.
(Courtesy of Ferragamo, Italy)

fashion, in out-of-the-way boutiques, or even in other countries. They are discerning shoppers who would rather have a few beautiful things than many mediocre things.

Most fashion leaders occupy prominent positions that give them exposure, and, in turn, influence over the way others dress. The press reports details of what they wear when they are seen at public events, in films, or on television. Because of their support, fashion leaders are a stimulant to designers and to the fashion industry as a whole.

There are fashion leaders for every market segment. Hillary Clinton, although she has many more important things to do than go shopping, was suddenly placed in the fashion limelight when her husband was elected president. She presents a very professional image that is a helpful role model for many women who identify with her. International fashion pacesetters include Diana, Princess of Wales, who is constantly photographed by the press. Many young women want to look like the celebrity models such as Claudia Schiffer and Cindy Crawford. For juniors, rock stars such as seen on MTV are the fashion leaders who set the trends.

Most fashion leaders are confident of their own taste and do not need the approval of others. As individualists, they do not need the security of standardization. They dare to be different, and their acceptance of a designer's clothes makes those clothes fashionable.

Fashion followers

Fashion needs followers or there wouldn't be any fashion. Of course, we should also consider *fashion victims*: those people with too much money to spend who become slaves to signatures. Designer Jean-Paul Gaultier remarked, "Fashion victims are people who blindly and stupidly follow a brand without any discernment and without any analysis. As long as it's the latest rage, they buy it without thinking about adapting it to themselves."[5]

Most men and women, however, seek acceptance through conformity and follow world, national, or community fashion leaders in order to feel confident. Fashion followers emulate others only after they are sure of fashion trends. Consumers become fashion followers for one or more of the following reasons:

◆ They lack the time, money, and interest to devote to fashion leadership. And, after all, most people have more important things to do.

◆ They need a period of exposure to new styles before accepting them.

◆ They are insecure about their tastes and therefore turn to what others have already approved as acceptable and appropriate.

◆ They want to keep up with their neighbors or peer group or be accepted by them.

◆ They tend to imitate people they admire.

Because of fashion followers, most members of the fashion industry are copyists or adapters. From a marketing point of view, fashion followers are very important. They make mass production successful, because volume mass production of fashion can only be profitable when the same merchandise is sold to many consumers.

ADOPTION OF FASHION

Basically, there are three variations of the fashion adoption process: traditional adoption, reverse adoption, and mass dissemination. It is important to understand how new fashion ideas are disseminated, or spread, and how they are adapted to the tastes, life-styles, and budgets of various consumers.

Traditional Fashion Adoption

Innovative designers have the courage and confidence to try new looks. At first the looks seem outrageous to many people, until it is seen how

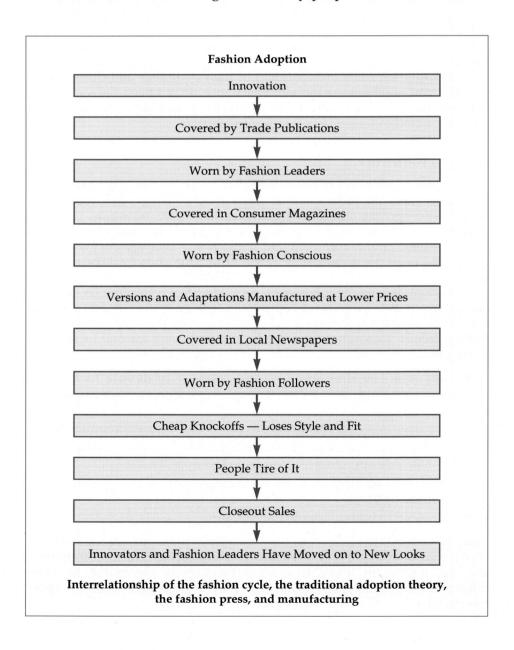

Fashion Adoption

Innovation
↓
Covered by Trade Publications
↓
Worn by Fashion Leaders
↓
Covered in Consumer Magazines
↓
Worn by Fashion Conscious
↓
Versions and Adaptations Manufactured at Lower Prices
↓
Covered in Local Newspapers
↓
Worn by Fashion Followers
↓
Cheap Knockoffs — Loses Style and Fit
↓
People Tire of It
↓
Closeout Sales
↓
Innovators and Fashion Leaders Have Moved on to New Looks

Interrelationship of the fashion cycle, the traditional adoption theory, the fashion press, and manufacturing

they can be adapted to individual life styles. To attract attention and change direction even moderately, high fashion often seeks to be extreme. Some film stars such as Cher or Madonna wear extreme fashion to get the attention of the press; however, most extreme designs are not accepted by the general public.

The traditional adoption theory is based on the fact that most high fashion is expensive, and therefore affordable to only a few people. As the new fashions are worn by publicized fashion leaders or shown in fashion publications, more consumers are exposed to the new look, and some will desire to have it for themselves. To appeal to this broader group of consumers, manufacturers produce less expensive versions or adaptations of high fashion. These are copied again and again at lower prices, until they have been seen often enough to become acceptable to the most conservative buyer. The cheapest versions are seen at discount houses soon after. Consumers then tire of the look and its popularity fades.

The length of this process is influenced by location. If the new look starts in Europe, then New York and other large cities will probably be the first to accept it. It may take a year or two for many Americans to fit even a modified version of the look into their life styles. Fashion implies newness and freshness. Yet as a fashion is copied, modified, and sold at lower and lower prices, it loses its newness, quality, and other essential design elements.

Reverse Adoption

Fashion occasionally begins with the consumer. In the 1960s designers began to watch people on the streets (street fashion) to find ideas. Some of these ideas eventually reach the designer market. For instance, the early 1990s saw designer alligator and leather "belt bags" from Ferragamo, Joan and David, Luc Benoit, and other manufacturers—direct descendants of the "fanny packs" (small nylon bags on a belt) originally worn by joggers. Today, designers pay more attention to consumer innovation. Joan Kaner, fashion director of Neiman Marcus, remarked, "The way [Karl Lagerfeld] translates what's going on in the street onto the runway is great."[6]

Mass Dissemination

Modern communication sometimes seems to make fashion available simultaneously throughout the industry. Such dissemination is evident when manufacturers copy hot new styles almost immediately in order to meet the high demand. Speed of production is of the greatest importance.

There is no longer one channel of fashion dissemination. Many separate markets have developed geared to various age ranges, life styles, tastes, and pocketbooks. Various designer and manufacturer labels appeal to various groups at different price points. Increased diversity means that many different styles can be acceptable at the same time. There are many more fashionable options.

Therefore, no single adoption theory is appropriate for the entire fashion industry. Missy market styling is generally adapted from designer fashion through the traditional process. Styling ideas for active sportswear and denims are usually inspired by the consumer, through the reverse process. Junior fashions might be mass-disseminated, since the original ideas are often produced precisely for this youthful market. We should also remember that many American consumers today are not even interested in fashion.

INDIVIDUALITY VERSUS CONFORMITY

We try to find a balance between individuality and conformity just as designers and merchandisers do when planning their products.

Most consumers seek varying amounts of both individuality and conformity in their dress. To feel a sense of belonging to their peer groups, they follow fashion; to express their personalities, they find ways to individualize fashion. A consumer can select from a wide variety of fashion looks on the market, combining and accessorizing them for self-expression, which is particularly important in today's mass-oriented society. Yet many consumers find the variety of choice confusing, and this makes their shopping decisions difficult. These people often gain support from knowing that others share their preferences.

The fashion industry reacts in the same way. Manufacturers and retailers try to establish a particular merchandising specialty, a uniqueness that their customers can identify with. Yet few manufacturers or retailers want to be outside the mainstream of fashion.

SUMMARY

Fashion has three properties: style, acceptance, and timeliness. Change makes the fashion world go round. New styles are introduced, rise to a height of popularity, then decline into obsolescence. Some styles remain fashionable longer than others. Some come back into fashion after a latent period. Because fashion is a product of change, a sense of timing is an important attribute at all levels of the industry.

Consumers can be identified with stages of the fashion cycle. Fashion leaders, a very small percentage of the population, buy and wear new styles at the beginning of the cycle; other consumers imitate. The majority of consumers are followers, which facilitates the mass marketing of fashion.

Fashions generally sift down from the original ideas of high-fashion designers, to reappear in cheaper fabrics at lower price levels. Other styles are either adapted "up from the streets" or are disseminated quickly through mass marketing. Consumers can choose from a wide variety of styles, dressing to conform to peer group preferences or to be individualistic. Consumer acceptance is a major influence on styling and merchandising decisions.

CHAPTER REVIEW

Terms and Concepts

Briefly identify and discuss the following terms and concepts:

1. Fashion
2. Style
3. Acceptance
4. Timeliness
5. Fashion evolution
6. Fashion cycles
7. Phases of fashion cycles

8. Classics
9. Fads
10. Fashion leaders
11. Fashion followers
12. Fashion adoption
13. Mass-market dissemination

Questions for Review

1. Name and define the three components of fashion.
2. How does fashion acceptance affect the timing of design?
3. Describe the phases of a typical fashion acceptance cycle.

4. Discuss the relationship of consumer acceptance to the fashion cycle.
5. Relate the various types of fashion adoption to various consumer groups.

Projects for Additional Learning

1. Is there a well-known movie star or public figure (either male or female) in the news whom you feel is a fashion leader? Write a short report documented with examples of how this person influences the way others dress. Clip (or photocopy) illustrations from publications.

2. From current fashion magazines, collect five examples of each of the following types of fashion: (a) high fashion; (b) mass fashion; (c) classic; and (d) fad.

NOTES

[1] Quoted in "Fall Fashions: Buying the Line," *Time*, April 23, 1984, p. 77.

[2] As quoted in "King Karl," *Women's Wear Daily*, November 20, 1991, p. 7.

[3] As quoted in "Is the Designer Dead?" *Women's Wear Daily*, April 7, 1992, p. 10.

[4] Ibid, p. 11.

[5]. As quoted in "Les Fashion Victims," *Women's Wear Daily*, January 21, 1992, p. 4.

[6] As quoted in "Hard Choices in a Difficult Season," *Women's Wear Daily*, March 22, 1993, p. 14.

Designer Jessica McClintock working with the fabrics, trims and books that inspire her.
(*Courtesy of Jessica McClintock, photograph by the author*)

4

FASHION RESEARCH AND ANALYSIS

CAREER FOCUS

Every executive, designer, and merchandiser in the fashion industry must be involved with research and analysis. They must constantly study the life styles of consumers; shop the market; read trend and design reports, fashion magazines, and newspapers in order to understand consumers and what they might want to buy. This research is absolutely necessary to make intelligent planning decisions regarding design, manufacturing, and sales.

Market research companies offer career opportunities for researchers and consultants who study and report on demographics and consumer buying habits. There are also interesting editorial opportunities with fashion publications in every merchandise category; in sales promotion and marketing; and in photography and art direction. However, there are very few job opportunities as museum curators of costume.

CHAPTER OBJECTIVES

After reading this chapter you should have attained competence in the following areas:
1. Understanding the importance of research
2. The ability to comprehend market studies and fashion forecasting
3. Knowledge of design resources

*I*t would seem a simple matter to make or sell what people want to buy. However, awareness, good research, planning, and perhaps a little intuition are needed for producers and retailers to make, buy, and sell what consumers will want. Without proper research, or with an unexpected turn of events, merchandise ends up on the sale racks, causing losses for producers, manufacturers, and retailers alike.

It would be impossible to ask all consumers what they will want to wear a year or two in advance—they would not even know themselves. Therefore, a designer or merchandiser must anticipate their wants and needs by being *aware* of what is going on in the world and how society will affect fashion. Sometimes this happens intuitively, but most often it must be carefully researched.

Research is done before any other design or merchandising activity, at every level of the industry. It becomes second nature to every decision maker in the field of fashion. Because the very term *fashion* implies a state of flux, textile producers, fashion manufacturers, and retailers all need to absorb a constant flow of information in order to anticipate change and consumer preferences. Designer Karl Lagerfeld has said, "I want to be informed, to know everything, see everything, read everything...You mix all that, then forget about it and do it your way."[1]

MARKET RESEARCH

Executives, designers, and merchandisers must constantly research the market, sales records, and consumers to learn buying habits and preferences.

Shopping the Market

To study what consumers want and need, designers and merchandisers often visit retail stores to see what is selling and to try to find out why. Designers compare the styling, price, fit, and quality of lines that compete with their own. They also analyze lines at higher price levels to see how their own designs measure up. Retailers observe competing stores to analyse which merchandise selections and presentation methods are the most successful.

Sales Records

Every manufacturer and retailer keeps records of all sales. This has become much easier with computers. Properly interpreted, this information indicates customer preference for certain garments or accessories over others. Rising sales show what fashion trends are developing; declining sales show what styles have passed their peak. Overall weak sales show that a style is not meeting consumer needs for fashion, quality, or fit.

This kind of research is particularly necessary for volume manufacturers who cannot afford to take risks. However, since fashion is constantly evolving at a rapid pace, sales statistics alone are not a reliable way to evaluate trends. Also, they do not show what merchandise is *missing* in the stores.

Consumer Surveys

Surveys and focus groups are excellent ways to discover consumer wants and needs. Methods of questioning consumers can be formal or informal. Market research companies make inquiries by telephone or hold consumer focus group meetings. These meetings are very successful as typical target customers are selected to meet with manufacturers and retailers. These consumers discuss the pros and cons of merchandise that is presented to them. The consumer reactions are compiled and tabulated to see what preferences or rejections are indicated. Manufacturers and retailers pay these research companies for the information they collect. Other surveys, such as mailed questionnaires, are made by publications, manufacturers, and retailers.

Informally, researchers can obtain information by simply asking customers in a store what they would like to buy, what styles they like that are currently available, and what merchandise they want but can't find. Because of their close contact with their customers, owners of small stores can often do this effectively. Experts point out that buyers should have their offices on the selling floor so they can interact with consumers and give feedback to the manufacturers. It will be helpful when a computer program is developed that lets consumers directly input information about what they cannot find in the stores.

FASHION FORECASTING

Because retailers want to buy the fashions that their customers will want, and since fashion manufacturers must work so far ahead of the selling season, both must learn to be fashion forecasters. Fashion forecasting involves the following:

◆ Studying market conditions: how the consumer's buying behavior is influenced by society, economics, technology, and the environment.

◆ Evaluating the popular designer collections to find fashions (colors, silhouettes, fabrications, lengths) that suggest new directions.

◆ Noting street fashions (what people in the street are wearing), and keeping up with current events, the arts, and the mood of the public in order to project new trends.

Evaluating the Collections

Traditionally, both manufacturers and retailers turn first to Paris for an indication of the newest fashion ideas. Each year, the members of both the couture and the prêt-à-porter show their collections in fall and in the early spring. From all over the world, retail buyers and fashion editors invade Paris. They also go to Milan, New York, and sometimes London and Tokyo. Editors and buyers (and designers, too, although they are not invited to the shows) try to analyze the collections for outstanding new ideas which might influence fashion change.

Several designers may use a similar fashion theme because they have used common sources of inspiration, That several designers simultaneously respond with like ideas to the same stimuli may indicate a fashion trend. Certain designers such as Karl Lagerfeld set trends because so many other designers and manufacturers are influenced by his designs. Sometimes it is not a designer's total idea but only the fabrication, a silhouette, or another design element that becomes a trend. Very often a new trend appears in small doses until it spreads to other collections.

Fashion forecasters look for the styles they feel are prophetic—fresh ideas that capture the mood of the times and signal a new trend. They base their judgment on experience; awareness of fashion cycles; and economic, social, artistic, and technological influences on the consumer. As the press notices these similarities between collections and highlights them, the media exposure helps to make the trends. Buyers edit the collections when they buy for their stores, thus reinforcing trends.

Evaluating the collections becomes one way a designer can research fashion trends as a basis for a new line. As designers are not invited to the shows, they must evaluate by shopping in Europe, or using design services, magazines, and newspapers.

Trends for Target Markets

Fashion forecasting is becoming much more difficult. Diverse consumer life styles create many separate market niches, each with its own trends: trendy, junior, professional, family-oriented, and so on. With today's segmented market, most often a trend is confined to a single market niche. However, occasionally a very popular trend such as animal prints or the use of gold fabrics at holiday time spreads to all markets. Designers and merchandisers must decide what looks best suit their customers based on age range, income level, life style, and fashion preferences.

TREND INFORMATION SOURCES

While they do research themselves, fashion professionals rely on information sources such as design, color, and video services; newsletters; books; magazines; and newspapers to make sure they don't miss anything important.

Even trendsetting designer Karl Lagerfeld has said, "...I cannot ignore trends."[2] Textile and fashion designers use video and printed sources (magazines, trade papers, and design services) to analyze trends because they want their designs to fit into the mainstream of fashion. In addition to supplying trend information, these publications are good sources of ideas. Designers collect pictures showing appealing ideas for color, line, fabric, and trimmings. One idea may inspire a whole line.

Fashion Services

Designers and merchandisers can do their own research or rely on the help of a fashion service. A wide range of reporting, forecasting, and consulting services are available to the trade only on a subscription and/or fee basis. Some of the major fashion services are Carlin International, Design Intelligence, Dominique Peclers, Here & There, Kaleidoscope, Nigel French, Promostyl, Stylists' Information Service, TFS-The Fashion Service, Trend Union, and the Tobé Report. They may provide collection reports, forecasting, slides, consulting, and even original designs.

Collection Reports

Fashion reports provide the most immediate in-depth source of information about the collections. The services work very quickly, getting their reports out within two weeks after the shows, whereas magazines take months. Fashion reports analyze the collections at more depth than newspapers or magazines are able to. They may include photographs and/or sketches, slides, fabric swatches, and descriptions. The collections can be so overwhelming in scope that even those who attend the shows may need assistance in sorting out ideas and identifying major trends.

A display of various design forecasts from Design Intelligence Ltd., London.
(Courtesy of the Fashionworks, photographed by the author)

Forecasts

Fashion services provide their viewpoints of fashion forecasting or how they expect fashion to evolve (they no longer wish to be called predictives as they feel they do not have crystal balls). The services send *trend books* to their clients that may include descriptions, sketches, fabric swatches, and color samples. Most offer information on both men's and women's wear. Some include children's wear and knits. They may project color or fabrication tendencies, forecast silhouette or other design elements, and indicate trends for particular target markets. Design Intelligence, for example, has a specific activewear book and a jeanswear book. Each service has its own special approach. Some services, such as Insights for footwear, cater to special niche markets.

Consulting

Many services now also offer personalized services to directly help their clients develop their product. This can be on an individual consultation basis or they may actually develop an entire line for a company. The fashion service Here & There advised Federated Department stores (the retail corporation that owns Bloomingdales) that young consumers are worried about global warming. Buoyed by Here & There's words, Federated introduced a line of lightweight clothing that could be worn through nine months of increasingly warm weather.

Color Services

Some services specialize in color forecasting. Fashion professionals and colorists meet at least twice a year to try to analyze color cycles and the natural evolution of color preferences. For example, as black was the "power color" of the eighties, we might assume that white, earth tones, and vegetable dye colors are important for the ecologically minded nineties. Color direction, like all fashion, is influenced by the environment, culture, the economy, the arts, and the seasons.

Twice a year, fashion and textile industry professionals pool their knowledge and analyze color direction two years before a retail-target selling season. In November 1993, for instance, color experts met at the Color Association in New York to discuss what colors would be worn in America in fall and winter of 1995–96. Forecasts, including yarn colors or swatches, are usually sent out by the color services in time for designers and merchandisers to plan their color stories and purchase fabrics.

Color projection services include The Color Association of the United States, The Color Box, The Color Marketing Group, Color Play, and Pat Tunsky, Inc. Fiber companies also supply color forecasts for their customers. This service is discussed in Chapter 6.

Video Services

Video is an ideal medium for fashion reporting. Videofashion Inc. sends subscribers video highlights of the collections. Vidcat provides designer profiles and highlights of retro-fashions.

Fashion channels QVC and the Home Shopping Network on Cable TV provide another opportunity to see various designer collections and retailer presentations (see Chapter 13).

The Color Association of the United States (CAUS) Women's Forecasting Committee at work selecting shades that they expect to be popular with Americans. *(Courtesy of the Color Association of the United States)*

Loose-Leaf and Newsletter Services

These services include newsletters and industry surveys that can aid the designer or buyer in forecasting or in finding ideas. Some of the major services are *Fashion Calendar* (listing dates of upcoming collection openings and market weeks), *John Naisbitt's Trend Letter, NAMSB News* (National Association of Men's Sportswear Buyers), *RTW* (Ready-to-Wear) *Review,* and *Inside Retailing.*

FG is a newsletter issued quarterly by The Fashion Group International for their members. The Fashion Group, headquartered in New York City, is a global nonprofit association of professionals in the fashion business. Their newsletters and meetings are a very good source of information on current developments in the fashion industry.

Directories and Reference Books

Many directories and references can help designers and retailers get the information they need. Fairchild has published the *Fashion Guide,* a directory of 10,000 industry listings including international designers, manufacturers, and buyers. In addition, there are the *Chain Store Guides* of department stores and specialty stores. To help the designer *source* (industry jargon, meaning to locate) fabrics, there are *Fabric Intelligence*, a world directory of fabrics; *Fabric Sourcebook*, a directory of resources available in the U.S.; and *The Agent*, a fabrics and supplies index.

Fashion Magazines and Newspapers

Fashion trend research also depends on a variety of trade and consumer publications. *Trade* magazines and newspapers are intended for people working in the fashion industry; *consumer* publications are created for the general public. Both manufacturers and retailers must read as many publications as possible, because each fashion journalist edits fashion trends from a different perspective.

Fashion editing

Ideally, the role of the fashion editor is to educate: to provide fashion information from all phases of the industry in all parts of the world, to make the industry (in trade publications) or the consumer aware of all that is available, and to help them make wise and suitable styling and/or buying decisions.

Fashion editors, together with journalists, stylists, and photographers, act as the eyes and ears of the consumer. They let the nation or the community know where to find the fashions that are currently on the market, and they report on how new fashions should be worn and accessorized.

The fashion editors of prominent newspapers and fashion magazines attend the collection openings, taking notes on what they like best and reporting on what directions they feel are important. They may request sketches or photographs of their favorite garments to keep on file. In between openings, fashion editors write articles on topics they feel are noteworthy. They sift through all the news releases that come into their offices to help them write a story. Editors may ask to borrow samples from manufacturers and have them photographed to illustrate the text or they may simply use a photo sent to them in a publicity release. Sometimes sketched illustrations are used, depending on the mood or effect to be achieved. Garment and fabric descriptions are included as well as a list of stores that carry the merchandise described in the article.

Trade publications

Trade publications in the form of newspapers, journals, or magazines offer information not only on fashion trends but on developments in the textile industry, fashion industry, and retailing business as well. Most specialize in a particular branch of the industry. Since busy executives have time to read only a few of these, they need to look for those that provide information relevant to their market. The following are some examples:

Women's wear: *Women's Wear Daily,* the major American fashion newspaper, is published five days a week. Each day features different topics and there are often special supplements on subjects such as accessories, sportswear, swimwear, or technology. *W*, a bimonthly condensed version of *Women's Wear Daily*, is produced in color but is full of advertisements. *California Apparel News*, published weekly, concentrates on the California market. *Style* is Canada's trade magazine, *Gap* (written only in French) is from France, and *Fashion Folio* is from England.

Men's wear: *DNR, Daily News Record*, another five-day newspaper with color supplements.

Children's wear: *Earnshaw's* and *Children's Business*.

Sportswear: *Sportwear International*.

Accessories: *Footwear News Magazine* and *Accessories Magazine*.

Textiles: *America's Textiles International, Bobbin*, and *Textile World*.

Apparel industry: *Apparel Industry Magazine* and *Apparel Manufacturer*.

Advertising and Marketing: *Advertising Age, Brandweek, American Demographics, Direct Marketing, Promo, Retail Ad-Week*, and *Sales & Marketing Management*.

Retailing: *Stores (National Retail Merchants Association), Chain Store Age Executive, Journal of Retailing*, and *Retail Store Image*.

Visual presentation: *Artweek, Communication Arts, Display & Design Ideas, Graphis, Inspiration, Lighting Dimensions, Views & Reviews*, and *Visual Merchandising & Store Design*.

International consumer fashion magazines

These publications are aimed at the consumer and show fashion that is already available in the retail stores. Therefore, Americans find foreign magazines much more helpful for discovering new trends because European fashion is more *avant garde* than American. It is not necessary to read the text of the foreign language magazines; the ideas come from the photographs. Some are published only two or four times a year to report on the collections. Many German and Japanese magazines have patterns included. In addition to the following list, there are many general interest magazines that include fashion.

Women's wear: *L'Officiel de la Couture et de la Mode de Paris, Jardin des Modes, Joyce*, and *Vogue* (France); *Line a Italiana, Donna, Moda, Harper's Bazaar Italia*, and *Vogue* (Italy); *Harper's and Queen*, and *British Vogue* (England); *Burda International, Madam, Fashion Guide*, and *Vogue* (Germany); *Élégance* (Switzerland); *Fashion Yearbook, Hi Fashion, Mode et Mode, Mrs.*, and *Madam* (Japan).

Contemporary and junior; trendy: *Elle, Dépêché Mode, La Mode en Peinture* (France); *Moda In, Moda Viva, Amica* (Italy); *Brigitte, Miss Vogue, Carina* (Germany); *Non-no, 25 ans*, and *McSister* (Japan).

Children's wear: *Vogue Bambini, Donna Bambini*, and *Moda Bimbi* (Italy); *Sesame* and *Baby Fashion* (Japan). (Junior fashions are also a good source of ideas for children's wear.)

Men's wear: *L'Officiel Hommes, Vogue Hommes* (France); *L'Uomo Vogue, Linea Italiana Uomo, Mondo Uomo*, and *Uomo Harper's Bazaar* (Italy); *Männer Vogue* (Germany); *Men's Club, Mr.*, and *Dansen* (Japan).

Bridal: *Mariages* (France) and *Sposa* (Italy).

American consumer fashion magazines

New magazines such as *Mirabella, Lears, Victoria, First* and *Sassy* are targeting specific market segments. In addition, many established magazines are undergoing extensive facelifts, and there is a continuing influx of foreign publishers and editors in the United States. American magazines as resources are best used for analyzing appropriate fashion for specific target markets, examining American designer collections, and seeing what fashion is currently available in the stores.

Women's: *Essence, Harper's Bazaar, Mirabella,* and *Vogue.*

Trendy: *Elle* (American Edition), *Glamour, In Fashion, Mademoiselle,* and *Sassy.*

Junior: *Seventeen* and *YM* (Young and Modern).

Men's: *Details, GQ (Gentleman's Quarterly), Esquire Gentleman,* and *EM (Ebony Men).*

Bridal: *Bride's, Modern Bride,* and *Bridal Guide.*

DESIGN SOURCES

Where do ideas come from? Why is it that so many designers appear to be doing the same looks at the same time? Ideas do not simply materialize out of thin air. Each designer strives to respond to his or her customers' needs. First the designer does careful research, but what makes a designer's line special is the unique interpretation of these influences.

Historic and Ethnic Costume

Designers often turn to the past (recent or distant) for line and silhouette ideas that can be used in a new way. They must become sensitive to the combinations of colors, motifs, lines, shapes, and spaces of design in each historical period of art and costume. Costume falls into two categories: *historic* costume, the fashion of a certain historical period, and *folk* or *ethnic* costume, traditional national or regional dress. Both are inspirational sources of design.

Historic inspiration

Faced with an uncertain future and discouraged by the reality of today, many designers retreat to the past. Karl Lagerfeld has recently been inspired by the 1970s. Fashion editor Maylou Luther remarked on the following influences in recent collections: "The romantic past of Catherine de Medici at Norma Kamali; Lillie Langtry and Anne Boleyn at Anna Sui, Josephine at Richard Tyler; the regal past of Czar Nicholas II at Ralph Lauren, the dandified past of Oscar Wilde at Anne Klein, and George Sand at Calvin Klein."[3]

Folk influences

Designers are often inspired by interest in other cultures. Byron Lars' fall 1993 collection had a Swiss Alps feeling. Christian Lacroix said that his

1993 spring collection "plays with mixes of South America, the Caribbean and Africa."[4] Donna Karan described her 1993 resort collection as, "...a mix, Peru, the Philippines, Mexico, all rolled into one."[5]

Museums

Designers find ideas in museums, books, and flea markets. Museum costume collections offer the unique opportunity of seeing actual preserved garments displayed on mannequins. Designers such as Yves Saint Laurent occassionally sponsor historic costume exhibits. A popular exhibit can influence many designers. The following museums and galleries house important costume collections:

Costume Institute, Metropolitan Museum of Art, 5th Avenue and 82nd Street, New York, New York 10028

Costume Gallery, Brooklyn Museum, 188 Eastern Parkway, Brooklyn, New York 11238

Costume Gallery, Los Angeles County Museum of Art, 5905 Wilshire Boulevard, Los Angeles, California 90036

Costume Institute, McCord Museum, 690 Sherbrooke Street West, Montreal, Canada

*Musée du Costume de la Ville de Paris,*14 avenue New York, 75016 Paris, France

Musée de la Mode, Pavillon de Marsan, Louvre, rue de Rivoli, 75001 Paris, France

Victoria and Albert Museum, Brompton Road, London S.W. 7, England (including costumes, old fabrics and embroideries)

Museum of Costume, Assembly Rooms, Alfred Street, Bath, England

Rijksmuseum, Stadhouderskade 42, 020 Amsterdam, the Netherlands

Kostümforschungs Institut (Costume Research Institute), Kemnatenstrasse 50, 8 Munich 19, Germany

Centro Internazionale Arti e del Costume, 3231 Palazzo Grassi, 30124 Venice, Italy

A suit from Byron Lars' Swiss inspired collection.
(Courtesy of Byron Lars, photo by Dan Lecca)

Many regional museums have fashion collections, and almost every national museum of folklore includes ethnic costumes. Costume can also be studied in paintings in every museum. The Matisse exhibit at the Metropolitan Museum in New York attracted visitors from around the world. One can see the Matisse influence in the Levi Strauss advertisements.

Costume can also be studied in books. Museum bookstores and libraries are excellent sources of costume references. Many libraries also have collections of old fashion magazines, including *Godey's Lady's Book* (from the 1830s to the 1890s), *Harper's Bazaar* (from 1867), and *Vogue* (from 1893). Many fashion designers have their own libraries of books on historic and ethnic costume, films, sports, artists, designers, and textiles. The rest are grateful for good public libraries.

An historic costume inspired design by Domenico Dolce & Stefano Gabbana for Complice.
(Courtesy of Genny U.S.A.)

The Arts

Film, video, television, art, architecture, music, and theater are also inspirational sources for fashion design. All designers are influenced by what other designers and artists are creating. Excitement about a new idea acts as a catalyst for more creativity.

Films (and therefore, their designers) occasionally influence style. The inspirations for Emporio Armani's *Casbah* look were three films: *Casablanca* (1942), *Laurence of Arabia* (1962), and *Aladdin* (1992).

Rock and rap music and the exposure that video gives singers and musicians has made a big impact on young people. The influence of the rap music culture can be seen in Cross Colours sportswear. The costume looks of stars as seen on MTV influence junior fashion.

All the arts give designers visual inspiration — for color, mood, or a certain spirit that seems to capture the times.

Fabrics, too, serve as inspiration for designers. Inspirational fabrics can be found in costume collections, in old or new interior decoration, or at the fabric market fairs. African tribal prints, for example, were inspirational to Scherrer and Valentino and used in their collections.

Travel

Travel can have a variety of influences on designers. Designers like to visit visually and culturally stimulating places. Travel exposes them to new ideas; however, these ideas aren't always obvious in their collections. Designer Mary Jane Marcasiano commented, "After going to Egypt, I got back into the earth tones and the practicality of natural fibers, natural colors, and layering."[6]

The Street Scene

Designers get ideas just watching people in the streets: people on their way to work, the trendy young girls, people who shop at thrift stores, the kids on their skate boards, or even the homeless.

Awareness

Designers surround themselves with photographs of ideas, fabric swatches, and anything else that will stimulate creativity. They leave their studios to shop, visit museums, study nature, attend the theater, or people-watch. Designers usually carry sketchbooks, to jot down ideas whenever and wherever they find them. They hunger for information, letting ideas mingle and shape themselves into new forms.

Awareness is the key to creativity. Designers must learn most of all to keep their eyes open, to absorb visual ideas, and to translate them into clothes their customers will like. Some people are more sensitive to good composition than others, but practice and observation make a person more aware, sensitive, and confident. Exposure to beautiful things helps a designer or buyer distinguish genuine beauty and quality from fads and mediocrity.

Designers and merchandisers must be aware of everything around them that affects the clothing industry, including economics, politics, demographics, and social change (see Chapter 2). It is impossible to predict where an idea will come from or which idea may inspire a whole group or line.

SUMMARY

Research and observation are critically important in the fashion business. By observing and researching consumer buying habits, watching for directions in designer collections, reading the most appropriate fashion publications, observing fashion trends, and being aware of the arts, manufacturers and retailers try to predict what the majority of their customers will want in the foreseeable future. To keep up with the changing world of fashion, fashion awareness should become second nature.

CHAPTER REVIEW

Terms and Concepts

Briefly identify and discuss the following terms and concepts:

1. Fashion editing
2. Fashion forecasting
3. Fashion trends
4. Target market

5. Evaluating collections
6. Fashion services
7. Trade publications
8. Consumer publications

Questions for Review

1. Why is awareness important for the designer and the buyer?
2. Explain fashion trends.
3. Explain the difference between trade and consumer fashion publications.
4. What role does historic or folk costume play in today's fashion?

5. Give an example of how films have influenced fashion.
6. Explain what research a designer or merchandiser needs to do.

Projects for Additional Learning

1. Trend research: Make a survey of fashion resources available to you. Read the latest issues of fashion magazines and compare the contents with fashion looks six months ago and a year ago. From the information you find, try to analyze the fashion direction for the future. Organize your ideas into a few basic trend forecasts for color, fabrics, silhouette, and line.
2. Fashion-magazine evaluation: Examine and compare the content of four different fashion magazines. How many pages are devoted to paid advertisements? How many pages are devoted to editorial fashion reporting? Which magazine has the most interesting information? Why?

3. Look through costume books for pictures of ethnic dress. Using these as inspiration, design a garment using the line, shapes, colors, and details from these garments in a new combination for today's life style.
4. Visit a local art museum. Select a favorite painting —period, contemporary, still life, landscape, or abstact—that inspires you because of its colors, lines, shapes, or any other details that you like. Try to use one of the elements as the basis for a garment design sketch. If it doesn't work out, try another painting until you feel you have achieved success.

NOTES

[1] As quoted in "King Karl," *Women's Wear Daily,* November 20, 1991, page 7.

[2] As quoted by Heidi Lender, "The Latest From Lagerfeld," *Women's Wear Daily,* February 9, 1993, p. 4.

[3] "Fashion Trend Report," The Fashion Group International, Inc., April 13, 1993, p. 1.

[4] As quoted in "The Paris Scoop," *Women's Wear Daily,* October 14, 1992, page 11.

[5] As quoted in "Resort," *Women's Wear Daily,* June 4, 1992, page 1.

[6] As quoted by Sara Fiedelholtz, "Sportswear Makers: Boosting Figures the Natural Way," *Women's Wear Daily Sportswear Trends Supplement* (February 1993), p. 2.

Part Two

THE RAW
MATERIALS
OF FASHION

A field of flax. *(Courtesy of the International Linen Promotion Commission)*

5

TEXTILE FIBER AND FABRIC PRODUCTION

CAREER FOCUS

Careers in textile production require a technical education in textiles and/or engineering. Since most mills in the United States are located in the Southeast, this is the area where most jobs are found. There are technical career opportunities in research and development, styling, plant technology, engineering, and management. Designers and manufacturers also need to understand textile production.

CHAPTER OBJECTIVES

After reading this chapter you should have attained competence in the following areas:
1. Awareness of the sources of fibers
2. Understanding of the processes involved in the production of fibers and fabrics
3. Knowledge of the roles of mills and converters

*B*efore we can even begin to think of a finished garment, we must consider the raw materials from which it is made. Collectively, the producers of these raw materials are the suppliers to the apparel industry.

Primary suppliers are the producers of textiles; furs and leathers; and of certain plastic, paper, and other nonwoven materials. Secondary suppliers provide the *trimmings* (decorative materials including buttons, laces, and ribbons) and *findings* (functional materials such as zippers, elastic, interfacing, and linings) needed to complete a garment as discussed in Chapter 7.

Since most garments are made from fabrics, the textile industry is our major concern. *Textiles* is a broad term referring to any material that can be made into fabric by any method. Occasionally the term *textile industry* is used to cover the whole apparel industry: the production and marketing of textile merchandise from raw materials to final product.

More precisely, the textile industry encompasses the production and marketing of fibers, yarns, and fabrics, including trimmings and findings. This production and marketing chain includes the following steps: fiber production, yarn production (spinning and texturing), fabric production, dyeing, printing, and finishing. Defined in this way, the textile industry represents the first level of the fashion industry.

TEXTILE PRODUCERS

Textile mills and converters are the producers of yarns and fabrics and the suppliers to the apparel industry.

Chemical Companies

Large chemical companies such as BASF (Badische Anilin und Soda Fabrik), Courtaulds Fibers, Inc., DuPont, Hoechst-Celanese, and ICI (Imperial Chemical Industries) pioneered the development of man-made fibers. The fiber divisions of these companies do extensive research and development of both fibers and fabrics and work closely with their customers, the textile mills, and provide services to manufacturers and retailers.

The current trend in the textile industry is toward consolidation and specialization. Robert Laforce, former manager of creative services at Hoechst Celanese Corporation commented, "Strong competitive pressures from imports have forced the U.S. textile industry to retrench and focus on those things they do best."[1] This specialization has led to international merging and acquisitions. Lenzing of Austria bought the rayon division of BASF. DuPont bought ICI's nylon facilities and, in turn, ICI purchased DuPont's acrylic division.

Textile Mills

Mills are the producers of yarns and fabrics. Some mills produce only yarns. Others knit and weave unfinished fabric, called *greige* (pronounced "gray") *goods,* from their own or purchased yarns. Many mills produce both greige goods and some finished fabrics. Greenwood, for example, produces both denims and greige goods, which it sells to *converters.* Small mills usually specialize in one type of fabric such as velveteen. Large companies create separate divisions to do the same thing, grouping similar fabrics under one division .

The various stages in textile production may or may not be handled by firms under common ownership. Firms have tended to grow either forward or backward into the production or marketing chain to form large *vertical mills,* which perform all processes, from top to bottom (fiber to fabric). Milliken and Burlington are examples of vertical mills.

A mill is committed to operating expensive machinery at full capacity at all times, so it must produce large quantities of greige goods to stay profitable. Because of the huge investment they have in machinery and technology and the expense of changing it, the large mills tend to serve the mass market with volume fabrics. Their commitments do not allow the flexibility of keeping up with fashion trends and the pressure for constant change.

Converters

Converters source (find the best quality goods at the best price) greige goods from domestic or worldwide mills and contract facilities to have them dyed, printed, and finished in accordance with the requests of their customers, the apparel manufacturers. Converters do not own any mills; they act as styling intermediaries between the mills and the manufacturers. Converters can be flexible and change products to keep up with changing fashion because they do not have commitments to keep their own machines in operation. Therefore, the converter is an important supplier of fashion-oriented fabrics to manufacturers. Examples of converters include Abraham, Brittany, Charter Fabrics, Erlanger, and Pressman-Gutman.

Modernizing to Survive

The American textile industry has been plagued with fierce competition from imports. Because of low-cost, third-world labor resulting in cheap fabrics from Asia, the market share for domestic fabrics has shrunk. As a result, the industry has been beset with buyouts, takeovers, consolidations, and closings. However, the optimistic people in the industry feel that consolidation was necessary to make the industry strong enough financially to purchase the technology required to become competitive with imports through capital-intensive (rather than labor intensive) production. Only the larger companies with large capital reserves are able to absorb costs for the new computerized equipment.

In domestic textile mills, everything is computerized: the spinning and texturing of fibers and yarns; high-speed weaving and knitting; and automated material handling, dyeing, and finishing.

Environmental Concerns

The growing demand of consumers for more environmentally friendly products has generated a new awareness and ingenuity on the part of the textile industry. The American Textile Manufacturers Institute has established an "Environmental Excellence" program to set up an awareness and action program. This includes environmental targets and audits to encourage recycling and environmentally efficient manufacturing and finishing processes.

New standards are being set for the growing, processing, printing, and dyeing of natural cotton fabrics. Sally Fox, president of Fox Fibers, has developed naturally colored brown, green, blue, and pink cotton that eliminates the need for dye. She explained, "I didn't invent them. They're native to the Americas and were grown 4,000 years ago. My contribution has been through improving the fiber so it could be machine spun."[2] New strains of cotton are being developed that are insect and water resistant and require little, if any, chemical insecticides or fertilizers. Organic farming techniques promise chemical-free cotton fibers. Various companies are working to develop nontoxic pesticides, environmentally friendly alternatives to PVA and finishing detergents; citric-acid cleaners to replace phosphate and chlorine, natural oils to supplant petroleum lubricants; bifunctional, low-impact dyes to decrease water use and chemical waste; resins to replace formaldehyde; and enzymes to replace acid washes. Unfortunately, some organizations within the textile industry are against the use of many of these environmentally friendly developments. In California, for example, the cotton industry has forced Fox Fibers to move out of the state.

The expense for all these improvements is the problem. The costs to American and European textile companies to maintain health, safety, and clean air and water are astronomical. It is therefore very difficult for environmentally responsible producers to compete with the low prices from mills in Taiwan and China where producers don't pay

A textile laboratory. (*Courtesy of the International Wool Secretariat, England*)

to clean up their environment. American textile manufacturers want to require imported textile products to be made under the same environmental, health, and safety standards to ensure fair competition as well as a clean environment throughout the world.

FIBER PRODUCTION

Before the textile industry can supply fabrics to apparel manufacturers, it must first develop and produce both natural and man-made fibers. Production is the process of generating a product such as fiber, yarn, or fabric.

Fibers are the hairlike raw materials that are spun into yarns and then made into fabrics. Total world fiber production is continually growing. Production is now 38 billion kilograms per year as compared to 22 billion in 1970 (only one half of worldwide fiber production is for apparel).[3]

There are over 6,000 textile plants dispersed throughout the United States and approximately one million people are employed by the domestic textile industry.[4] U.S. textile production is centered in the Southeast, especially in North and South Carolina, where overall expenses, especially labor, tend to be lowest. Because New York City is the nation's fashion capital, however, most textile companies have offices there, as well as sales representatives all over the country.

The vast differences between the *natural* and *man-made* fiber industries have resulted in markedly different operational and organizational forms, even though the final goal of both groups is the same—to produce fibers that fill consumer needs.

The natural fiber industry is dependent on animals and living plants. Thus, the production of natural fibers depends on climate and geography, and in most countries the majority of crops are produced by thousands of small farmers.

NATURAL FIBERS

Used for thousands of years, natural fibers are derived from either animals or vegetables.

The most important *natural fibers* are flax, wool, cotton, and silk. Others include ramie, jute, sisal, and hemp. There is a trend toward blending natural fibers to create new textures.

Flax

Flax is made from the fibrous material in the stem of the flax plant. This substance is processed into a fiber called *linen*. Flax is harvested by pulling up the plants in order to preserve the full length of the fibers. Extracting the fibers from the flax plant is a lengthy and complex process. Flax seeds

The scutching process. *(Courtesy of the International Linen Promotion Commission)*

and adhesive substances that bind the fibers together must be removed by a process called *retting.* Then, in the *scutching* process, the fibers are separated from the outer bark and the woody inner core of the stem. Further processing can include *hacking* or *combing* and drawing the flax into a continuous ribbon of parallel fibers ready for spinning into yarn.

Linen is the oldest known textile, dating back as far as the Stone Age. At one time linen was used extensively for bedding; this explains why we still call sheets, towels, and tablecloths collectively linens. Flax was an important crop in the United States until the invention of the cotton gin in 1792 made cotton cheaper to produce. Today, flax makes up only 2 percent of world fiber production or approximately 600 million kilograms annually. Eighty percent of the world's flax is grown in Russia; France is the largest producer of flax in the Western world. Because the United States no longer grows flax, there are no quotas on its importation.

The cool, crisp, lightweight qualities of linen make it especially suited for summer clothing. Renewed interest in texture and natural fibers has repopularized linen and linen blends as well as linen look-alikes in the form of rayon and other fiber blends.

Ramie

Ramie is a formerly rare fiber that is growing in popularity because its importation into the United States is not restricted as there are no domestic commercial growers. A vegetable fiber similar to flax, it comes from the five- to six-foot stems of a nettlelike shrub. Ramie grows best in a semitropical climate, and is imported primarily from India, China, and the Philippines.

Ramie is even stronger than flax and has a smooth, lustrous appearance. The fibers dye easily but are brittle and more difficult to spin and weave than other fibers. Therefore, it is most often combined with cotton to soften it.

Cotton

Cotton has long been the world's major textile fiber. It comprises about one half of world fiber production or approximately 22 billion kilograms annually, but this amount fluctuates. A vegetable fiber, it grows best in tropical and subtropical climates. China grows the most cotton, followed by the United States and the southern republics of the former Soviet Union such as Kazakhstan.[5] Smaller producers are India and Pakistan. In the United States, fourteen states in the South make up what is known as the Cotton Belt, with Texas having the largest acreage under production. The Cotton Belt stretches from the Southeast through the Mississippi Delta to Arizona and California.

A ripe cotton boll. *(Courtesy of the Cotton Council)*

The cotton plant has blossoms that wither and fall off, leaving green pods called *bolls.* Inside each boll, moist fibers push out from newly formed seeds. The boll ripens and splits apart, causing the fluffy cotton fibers to burst forth. The cotton is then picked and *ginned,* an operation that separates the fiber from the seed. The fibers are further cleaned and also straightened by the *carding* process.

Cotton fiber may be processed on a combing machine that removes seeds, short fibers, and neps, resulting in a smooth, uniform yarn. The quality of cotton depends on the length and fineness of the fiber or staple, long-staple cottons being the best quality. Among American cottons, Sea Island is considered the finest, followed by pima.

Cotton is washable and durable, holding up well after many launderings. But because it has no elasticity, it wrinkles easily (however,

Cotton harvesting. *(Courtesy of the Cotton Council)*

wrinkle-resistant finishes have recently been created that make cotton easier to care for). Cotton absorbs dyestuffs easily to produce a wide range of vivid colors. It also absorbs moisture, which makes it feel cool against the skin in hot, humid weather. For that reason cotton has traditionally been a summer fabric. Yet cotton is very versatile and can be made in both light weights for summer and heavier weights for winter. Cotton fabrics range from the light and sheer (such as voile and batiste) to the heavy and thick (corduroy, flannel, and chenille) to the strong and sturdy (denim).

Wool

Wool fiber comes from the fleece of animals, most commonly sheep. The *fleece* is removed by the use of shears or clippers in the *shearing* process. Wool fleece is a natural and renewable resource: after a year the wool has completely grown back and is again ready for shearing.

Although wool is commonly understood to be fiber from the fleece of sheep—most wool in clothing is sheep's wool—some other animal fibers, usually called specialty hair fibers, are also classified as wool. They include camel's hair, cashmere, mohair, and alpaca.

Wool fiber has an unusual ability to absorb and evaporate moisture. The crimp structure of wool fibers allows it to return to its natural position after stretching, which gives it resiliency. When woven or knitted, the crimp structure creates air pockets which give it insulating properties. Because it is warm, wool has traditionally been used for fall and winter suits and coats. Yet it can also be woven or knitted into lightweight fabrics such as challis. Wool fibers are firm, yet soft and resilient, making wool fabrics resistant to wrinkling and comfortable to wear. New developments in wool fabrics include blends of cashmere, wool and angora; metallic sparked suitings; and lightweight velour fleece.

Raw wool is *carded* (cleaning the fibers by separating and laying parallel)and *combed* to separate short fibers from long ones. The long fibers are spun into smooth, compact *worsted* yarns used for fabrics such as gabardine or crepe; short fibers make up the soft, dense *woolen* yarns used in textured tweeds and flannel.

Australia is by far the world's largest wool producer, followed by the southern republics of the former Soviet Union such as Kazakhstan and New Zealand. Although more than 1.9 billion kilograms of clean wool are processed each year, this represents only 5 percent of world textile fiber production.[6] The supply of wool changes yearly; sometimes there is a stockpile and other times a shortage.

Shearing Merino sheep in Australia.
(Courtesy of the Wool Bureau)

Silk

Silk is the only natural fiber that comes in continuous *filament* form as opposed to short staple lengths. *Silk* is the protein filament secreted by a silkworm to make its cocoon. The silkworm, the forerunner of the silk moth, uses the cocoon as a shell to protect itself during its transformation from caterpillar to moth. Silk harvesters unwind the filament from the cocoon onto silk reels. A typical cocoon will produce 600 to 2000 meters of continuous fiber, or over 500 silkworm cocoons to make a blouse. The time and labor needed to cultivate silk makes it a rare and therefore expensive fiber. World production of silk fiber is a mere 0.2 percent or only 60 million kilograms annually. Asia produces 95 percent of the world's raw silk filament fiber. Asia has decreased exports of raw silk and increased production and exports of finished silk fabrics. The most famous quality silk fabric mills are in Como, Italy, but the Koreans are learning their techniques.

There are four kinds of silk fibers: *Cultivated silk* comes from the domesticated silkworm. The filaments are almost even in size, their fineness indicated by a unit called the *denier*. Cultivated silk is used for the finest silk fabrics, such as crepes, taffetas, and satins. *Wild* or *tussah silk* comes from the wild silkworm. Less secure environmental conditions cause the filaments to be coarser and more uneven. Therefore, fabric made from wild silk is not as smooth as that made from domesticated silk. *Douppioni silk* is the filament from two or more cocoons that have grown together so that the fibers join at intervals. Yarns made from these fibers have thick, uneven nubs from the joinings. Such yarns are used in shantung (a rough-textured silk made from uneven yarns). *Schappe* and *bourette (waste) silk* are composed of short fibers from damaged cocoons, not strong or long enough to be used on their own. Yarn spun from waste silk also has irregular slubs (uneven yarns); it is used in rough-textured silks.

Silk has always been used for the finest garments. The silk fiber is triangular and reflects light, giving silk its unique luster. Because it takes dyes with exceptional depth and clarity and has a luxurious feel, it adds elegance to any garment. Silk also has insulation properties, making the wearer feel cool in summer and warm in winter. Silk drapes exceptionally well, is very strong yet lightweight, and is comfortable as well as beautiful.

MAN-MADE FIBERS

The phenomenal growth of the textile fiber industry would have been impossible without the development of man-made fibers.

Chemists began to experiment with synthetic fibers as early as 1850. Originally the term *synthetics* was used to denote all chemically produced fibers. Today, however, the textile industry has chosen the term *man-made* (perhaps a poor choice linguistically). In the world market the term *synthetic* covers only the noncellulosic fibers. A true synthetic is made from the carbon atom of oil or coal, while viscose and acetate are rejuvenated cellulose. Perhaps a better term for man-made fibers would be simply *manufactured* fibers.

In 1884 a Frenchman named Hilaire de Chardonnet patented a fabric he called "artificial silk," which is known today as rayon. He discovered that silkworms use cellulose from mulberry leaves to make real silk, and he reproduced the process chemically. Rayon was first produced chemically in the United States in 1910, but not until 1939 was the first completely chemical fiber, nylon, introduced by DuPont.

Man-made fiber production depends mainly on the supply of petroleum products; therefore, it is subject to shortages and ever-increasing prices. In contrast with natural fibers, man-made fibers are produced by a few chemical companies, whose huge facilities enable them to take full advantage of mass production. These factories are located in areas where the combined cost of raw materials, labor, energy, and transportation is the lowest possible.

The growth rate of man-made fibers far outstrips that of natural fiber production. In 1960, man-made fiber production was only 3.511 billion kilograms or 22 percent of world fiber production, as compared to over 17.6 billion kilograms or 46 percent of world production today (however, only about one third of this amount is used for apparel).[7] This growth is due to many factors: fiber research and development by chemical companies, new technical developments in making new varieties of fibers, the scarcity and expense of some natural fibers, and increased production facilities in Asia. The United States produces less than one quarter of that amount, having lost some of its share of the market to foreign competition. The United States, as a single nation, leads the world in man-made fiber production, followed by Taiwan, Japan, Korea, and China. However, if you combine the Asian countries, their output far exceeds that of the United States. Asia and the United States are followed by Western Europe (primarily Germany and Italy) and the former countries of the Soviet Union (Russia, Poland, and Kazakhstan).[8]

Of utmost importance is the use of man-made fibers in blends, either with each other or with natural fibers. Blends capitalize on the best qualities of each fiber. For example, a polyester might be blended with cotton to add easy-care properties to the look and feel of the natural cotton. Polyester might also be blended with rayon to create a less expensive substitute for flax.

There are two categories of man-made fibers: *cellulosic* and *noncellulosic* (synthetic).

Cellulosic Fibers

Derived chiefly from spruce pulp and other softwoods, cellulosic fibers create a *hand* (the feel, body, and fall of a fabric) more like that produced by natural fibers compared to chemical noncellulosic fibers. They include rayon, acetate, and triacetate.

Rayon

The first man-made fiber to be developed, *rayon* is composed of rejuvenated cellulose derived from wood pulp, cotton linters, or other vegetable matter and dissolved into a viscose spinning solution. Rayon has become very popular because it is a soft, lustrous, versatile fiber that can be treated and finished to produce a wide range of characteristics. There are new developments of silklike rayon microfibers. However, soaring wood prices have made rayon expensive and the caustic materials needed for the wet-spinning process cause pollution.

Lyocell (CF0001)

Lyocell is the generic designation in the United Kingdom for a new type of solvent-spun cellulosic fiber developed by Courtaulds (a chemical/fiber company), known commercially as Tencel. In the United States it is currently known generically as CF0001 until the Federal Trade Commission gives it a generic name. Lyocell is twice as strong as rayon, both wet and dry, and is a good blending partner with cotton, wool, and polyester. It takes color well and its hand can be altered to resemble silk or cotton. The wood pulp is provided by eucalyptus trees that are specifically cultivated for this purpose. Another positive attribute is that the lyocell solvent spinning process is nonpolluting because the solvents are continually recycled.

Acetate and triacetate

Acetate (diacetate) and *triacetate* have been adopted as less expensive, less polluting (they can be dry-spun) alternatives to rayon as they are also soft and lustrous, although not quite as strong as rayon. Both are cellulose chains treated with acetic anhydride. However, only two thirds of the chain is treated to produce diacetate. The entire chain (all three parts) is treated to produce triacetate. Then both kinds (acetate and triacetate) are dissolved in a solvent to make dry spinning possible. Triacetate can be heat-set to give it easy-care properties, but it requires such strong solvents that it is no longer produced in the United States. Diacetate can be dissolved in solvents that are easier to handle and therefore is more widely produced.

Noncellulosic or Synthetic Fibers

Made from chemical derivatives of petroleum, coal, and natural gas, synthetic fibers used in apparel include nylon, polyester, acrylic, spandex, and polypropylene.

Nylon

One of the most durable man-made fibers in spite of its extreme light weight, *nylon* is made of a long-chain synthetic polymer that has recurring amide groups. Scientists have learned to vary the shape and width to

Nylon's light weight but warm properties make it ideal for ski wear.
(Courtesy of Fila Sports Inc.)

develop nylons with a crisp, even, silky hand. The filaments can also be textured to increase their suppleness and bulk. Nylon is not only strong but also flexible, washable, and colorfast. In apparel it is used primarily in hosiery, lingerie, bathing suits, and active sportswear.

Polyester

A long-chain synthetic polymer (composed primarily of dihydric alcohol ester and terephthalic acid), *polyester* is the most widely used man-made fiber in the world. Fiber forms produced are filament, staple, and tow (short or broken fiber). The production process resembles that of nylon. There is a wide variety of polyester fibers that differ in, among other variables, their shape, whiteness, type of crimp, and whether they are solid or hollow. Polyester is highly wrinkle resistant and easy to care for. It was one of the first fibers to be developed into fabrics with permanent-press features. It is often blended with natural fibers to lend easy-care properties to them. New developments include microdenier polyester. It is used in many types of apparel including textured knits and wovens, permanent-press blend fabrics, shirtings, suitings, and sleepwear.

Acrylic

A long-chain synthetic polymer (composed mainly of acrylonitrile units), *acrylic* fiber is thermoplastic, allowing fabrics to be heat-set for wrinkle resistance and permanent pleating. Because of the composition and cross-section of the fiber, fabrics made of acrylic have a high bulk-to-weight ratio. This provides warmth in fabrics that are lightweight, soft, and resilient. End uses of acrylics include knitwear, fleece activewear, suits, and coats.

Spandex

A long-chain synthetic polymer comprised of segmented polyurethane, *spandex* can stretch over 500 percent without breaking and return to its original length. Its elastic properties are unequaled by any other fiber and it does not deteriorate as rubber does. Because of its tremendous stretch, spandex is generally blended with other fibers in small percentages of 2 percent to 20 percent. When blended, spandex is essentially invisible. This characteristic makes it an excellent fiber for use in swimwear, hosiery, and active sportswear. Spandex is very popular and is blended with cotton, nylon, rayon, and even flax to create stretch linen. It is used in stretch wovens as well as knits and even stretch satin for evening wear! It is most often seen advertised under the DuPont brand name of Lycra.

Polypropylene

An olefin fiber made from polymers, *polypropylene* is very strong and resilient. It provides greater coverage per pound than any other fiber yet it is so light that it actually floats. While major uses of polypropylene are industrial and carpet applications, its good insulative properties make it desirable in high-tech activewear where moisture transport is important.

Generic and Brand Identification

When a completely new fiber is developed, the United States Federal Trade Commission assigns it a generic name, such as *polyester.* Many companies produce polyester but they each call it by a different name in order to promote it independently. A *brand* name or *trademark* (identifiable symbol) is registered and is owned by the manufacturer. For example, *Dacron* is DuPont's brand name for polyester, while *Trivera* is Hoechst-Celanese's brand name. The Textile Fibers Products Identification Act requires that end products carry labels listing the generic names of the fibers used and the percentage of each.

It is possible to work within the basic generic composition of an existing generic fiber and modify it, both chemically and physically, to change the properties of the fiber to create variants. *Variants* are specialty fibers for special applications. For example, Hoechst-Celanese created a silklike, lustrous polyester they call *Ceylon* to use as a silk substitute. Another variant called *Linenesque* was created with a new texture suitable for blending with rayon to imitate linen. Not all variants are intended to imitate natural fibers. Some are blended with natural fibers to add easy-care properties to them. Others are created to add stretch, moisture wicking, or other comfort properties. *Thermax* is DuPont's brand name for its insulating hollow-core polyester. There are also variants of other man-made fibers.

Man-Made Fiber Production

In very simple terms, all man-made fibers are extruded from a viscous solution of *cellulose* (purified wood pulp) or from chemical raw material. Originally the chemical substances exist as solids and therefore must be first converted into a liquid state. The raw materials are converted into flakes, chips, crumbs, or pellets, which are then dissolved in a solvent, melted with heat, or chemically converted into a syrupy liquid and pumped through the tiny holes of a spinneret to form continuous filaments.

The process of extrusion and hardening is called *spinning*, not to be confused with the textile yarn operation of the same name. There are three methods of spinning man-made fibers: wet, dry, and melt spinning:

Wet spinning As the filaments emerge from the spinneret, they pass directly into a chemical bath where they are solidified. Acrylic and rayon are produced in this way.

Dry spinning As the filaments come from the spinneret, they are solidified in warm air. This process is used in the production of acetate, triacetate, acrylic, and spandex.

Melt spinning This method is used for substances that were melted for extrusion and are then hardened again by cooling. Nylon, polyester, and polypropylene are produced by this process.

Unlike natural fibers, the man-made fibers can be extruded in different thicknesses, called *denier*. These long fibers can be left as continuous filaments, or cut into *staple* (short, uniform lengths) to be blended with other fibers.

Man-made fibers were first created to imitate the texture of natural fibers, duplicating their crimp, length, and thickness. Since then, scientists have learned how to vary the shape, composition, and size of fibers to achieve pleasing aesthetic effects and higher levels of performance.

The newest man-made fiber development is *microfibers* which are less than one denier thick, twice as fine as silk. Used primarily in polyester, this luxury fiber is giving polyester a renaissance. It can be used alone or combined with silk, worsted wool, or other fibers.

YARN PRODUCTION

Once a fiber is produced, yarn production is the next step in creating a fabric.

Textile manufacturers choose a yarn-making technique that can best achieve the texture, hand, or drape desired. Yarn can be made as coarse as rug yarn or finer than sewing thread.

Filament Yarn Processing

A *filament* is a continuous strand of fiber. Single monofilament, direct from the fiber producer, may be used as a yarn and knitted or woven into fabric. Multifilaments may also be brought together, with or without a twist, to create yarn. These yarns are smoother, have more sheen, and are more uniform than spun yarns.

Silk filament is twisted in a specialized silk spinning process. It is then ready for weaving or knitting.

Synthetic filament, on the other hand, is generally textured to provide bulk, loft, or elasticity. *Texturing* is a process used on filament yarn to change the shape or characteristics into some form of crimp, loop, curl, or coil. The two most widely used texturing methods are air-jet texturing and false-twist texturing.

With the *air-jet method*, the filament is overfed into an air jet, which loops and bulks the yarn by means of the turbulence. The advantage of this method is that it gives the filament a more natural hand.

The *false-twist method* is the most popular and least expensive method of filament texturing. The filament is twisted on a spindle, heat is applied to set the twist, and then the twist is removed with the yarn retaining a memory for the three-dimensional helical curl. After texturing, man-made filament yarn is ready for dyeing and/or weaving or knitting.

Spinning Staple Fibers

Spun yarns can be made from natural fibers and man-made fibers that have been cut into *staple* (short lengths that emulate raw natural fiber).

Natural fibers, such as cotton, flax, and wool, must go through a long and expensive series of processes to become yarn. Each fiber has its own special spinning system: the linen system, the cotton system, the wool system, or the worsted system. To explain the systems in a very simple way, there are first several processes to clean, refine, parallel, and blend the raw fibers. Next, the fibers are drawn out into a fine strand and twisted; this keeps them together and gives them strength to withstand the spinning process.

Man-made fibers are cut into staple and spun on the same conventional spinning system used for cotton or on a new high-speed system. The resultant yarns have characteristics similar to those of spun natural fiber. In fact, most man-made fibers that are cut into staple are used to blend with natural fibers. The short fibers are spun to join them together to make yarn. The blending of both natural and man-made fibers is very important as not all the fibers have the same quality. Blending gives the resultant yarn consistency.

Spinning systems

Ring spinning is the conventional method used for natural fibers whereby the fibers are spun and wrapped on a bobbin spinning in the middle of a ring. Because there is a metal bobbin in contact with a metal ring, it heats up and burns out if run too fast.

Open-end spinning has surpassed ring spinning in the United States because it is much faster. With this method, the fibers are twisted into yarn by the means of friction as they are flung against the sides of a hollow, high-speed turbine. This method is most suitable for heavier-weight cotton and man-made blended yarns.

Air-jet spinning is a new method developed by the Japanese that offers even greater productivity. By this means, the twist is obtained in an air vortex where the air drives the fibers around each other to form the yarn. Air-jet spinning allows the original character and softness of the fibers to carry over to the yarn. It is especially suitable for fine cottons and synthetics.

FABRIC PRODUCTION

After spinning, the yarns are ready for dyeing and/or weaving or knitting.

Fabric is material or cloth made from natural or man-made yarns by any of the following methods: weaving, knitting, bonding, crocheting, felting, knotting, or laminating. Most apparel fabrics are woven or knitted. Fashion preference for one or the other is usually cyclical.

Italy and Japan produce the most woven woolen apparel fabrics and Italy the most woolen knitwear. Better cottons come from Italy and

Warped yarns in preparation for weaving. *(Courtesy of the Cotton Council)*

Switzerland. However, the largest cotton fabric producer is China, followed by Eastern Europe, and then India and the United States. The United States produces more denim than any other country. Linen fabrics are produced in Italy, Belgium, Northern Ireland, France, China, and Poland. Italy produces the most printed silks but only one third of the total world silk fabrics. The other two-thirds come from Korea. It is difficult to calculate the production of fabrics from man-made fibers as so many are blends and most industry and trade associations calculate production only in their own countries. The three major global centers of man-made and blend fabrics are Asia, Europe, and the United States.

Weaving

Woven fabrics are made by interlacing *warp* yarn (lengthwise) and *filling* yarn (crosswise—called *weft* in England and in handweaving) at right angles. Weaving begins with a process called *warping*: yarns are wound onto a beam and hung lengthwise on the loom.

Industrial looms are enormous, accommodating tons of yarn and turning out huge quantities of fabric. There may be as many as 15,000 warp threads on one loom (ready-to-wear manufacturers need wide fabrics for efficient pattern layout).

In the conventional method of weaving, filling yarns are fed into the loom by a *shuttle* carrying yarn wound on a bobbin. The warp yarns separate alternately to allow the filling yarns to interlace with them as the shuttle passes through the warp shed.

A faster method of weaving uses *a shuttleless loom* that carries the filling yarns through the warp on steel bands. Still other types of shuttleless looms carry the filling yarns on tiny darts or tiny jets of water or air. These looms operate at three times the speed of shuttle looms and produce seven to eight times the fabric because they weave wider widths.

Weaving cloth on an automatic loom.
(Courtesy of the International Wool Secretariat, England)

Kinds of weaves

There are three basic weaves: plain, twill, and satin.

Plain weave is the simplest, most common weave. The warp and filling yarns alternately pass over and under each other, creating both horizontal and vertical surface interest. Fabrics woven by this method include a wide range of weights from broadcloth to duck.

Twill weave is created by passing the warp yarn over a number of filling yarns before going under one. The same pattern is repeated row after row, but each time the repetition begins on the next warp yarn, creating a diagonal weave that gives the cloth added strength and a diagonal surface interest. Denim is the most popular twill fabric.

Satin weave is achieved by one warp yarn crossing over the most possible filling yarns (or vice versa), creating floats on the face side of the fabric. The floats give the fabric luster and smoothness. But because they are caught into the fabric only at comparatively wide intervals, a satin weave does not have the same durability as the other weaves.

There are many variations of these three basic weaves used to create unusual designs or patterns in fabrics which are called *novelty weaves.*

Pattern created by weaving

Pattern can be introduced by using yarns of different colors in the warp, the filling, or both, to create plaids, checks, or stripes. These *yarn-dye patterns* can be distinguished from prints because the patterns appear on both sides of the fabric. Woven patterns may also be produced by reversing the direction of the weave in certain areas or in alternate rows, as in herringbones. Small figures, called *dobbies,* can be woven into the fabric by a special attachment on the loom.

Fancy woven patterns, such as brocade, damask, and tapestry, may be created on a *Jacquard* (jah-kard´) loom, run by computers. The pattern is programmed on a computer, which causes the individual warp yarns to go up or down, creating the desired pattern.

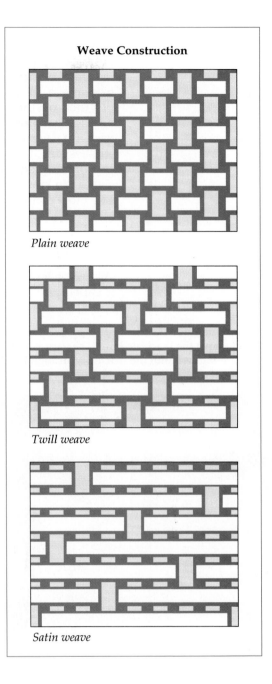

Weave Construction

Plain weave

Twill weave

Satin weave

Knitting

Knitted fabrics are made from one continuous yarn or combination of yarns formed into successive rows of loops drawn through another series of loops to make cloth. The amount of stretch in the final product depends on the fiber and yarn construction as well as the kind and density of the knit. Various types of stitches in combination with several ways of knitting produce a wide variety of knit fabrics from very bulky to fine-gauge knit jersey. On electronic knitting machines, designs can be transferred to a computer tape, which operates the machine automatically. A modern knitting machine can knit 1,000,000 loops a minute.

Gauge or *cut* refers to the number of needles per inch in the knitting machine; a 10-cut machine has 10 needles per inch. The number of stitches per inch on the finished fabric may be more or less than the number of needles per inch. A 5-cut machine with 5 needles is the most commonly used for sweaters. However, the resulting sweater can have as little as 3 stitches per inch or as many as 9 stitches depending on the amount and size of yarn fed into the machine. The more needles there are, the finer and closer the knit loops are. A fine jersey might be made on a 28-cut machine. The most common cut-and-sew jerseys and double knits are 18 to 24 cut.

Weft and *warp* knitting are the two basic methods of knitting fabric.

Weft knitting

When the loops run horizontally across the width of the fabric, the process is called *weft knitting.* Weft knits can be made on either flatbed or circular

A computerized circular knitting machine. The creel above the machine holds the yarn which is threaded down into the machine, through needles that knit the fabric. *(Courtesy of Cotton Inc.)*

machines, and circular knits may be either single or double knit. In single knits, all of the stitches in a given course or row are made with a single yarn. In double knits, the stitches in a given row may be made by interlocking two different yarn feeds. Weft knits are made in a wide variety of single and double knits and generally have more stretch than warp knits.

Jersey is the basic construction on all weft knits. The front and the back of the garment are different in appearance. The rows of stitches running across the garment are called *courses,* and the ones running vertically are called *wales.* Jersey is the most common stitch used in fine-gauge fabrics and stretches equally in length and width. Popular uses include pantyhose, underwear, and full-fashioned sweaters (see Chapter 10 for sweater and knitwear production). There are many variations on the jersey construction that are made by differing needle arrangements.

1. **Purl** knit is actually the reverse side of jersey and used primarily for sweater knits.
2. **Rib** knit fabrics have a distinctive lengthwise rib on both sides of the fabric for added stretch in the width. A combination of jersey and purl stitches, it is most often used in sweaters, socks, knit accessories, or trims such as neckbands or cuffs.
3. **Interlock** knit looks like jersey on both sides of the fabric. The stitches alternate a half movement up or down to create a zigzag horizontally across the fabric.

4. **Knit and welt** uses the front and back beds of the knitting machine to create a welt.
5. **Float jacquard** knits have a pattern on the face side. Yarns not being used on the pattern float on the back until they are needed again.
6. **Full jacquard** knits also have a pattern on the face side but there is another simple pattern on the back instead of floats. This requires the use of both the front and back knitting beds and makes a double fabric, which is heavier.
7. **Novelties,** including *tuck* stitches, *miss* stitches, and *pointelles,* can be created with other needle arrangements.

Warp knitting

In *warp knitting,* multiple yarns are used and the loops run vertically and zigzag across each other to form the fabric. Each stitch in a row is made by a different yarn that is fed from a sheet of yarns wound on a beam. Patterns and inlays can be introduced by various needle arrangements. The Missonis in Italy are famous for their beautiful warp knits. Warp knits include tricot and raschel.

Tricot (tree-co´) knit fabrics are usually made with fine-denier filament yarns. They are soft, drape well, and are somewhat elastic. Popular uses include women's lingerie and uniforms for nurses.

Raschel (ra-shell´) is the most complex warp-knit machine capable of making lacy open stitches. Yarns may be heavily twisted filament yarns, or spun yarns. Uses of this knit include thermal underwear, knitted lace, and crochet.

Nonwoven Fabrics

Nonwoven (or *engineered)* fabrics are made by either matting, bonding or interlocking fibers, filaments, or yarns into a web or sheet by mechanical (pressure, needle punch, or needle tufting), chemical, thermal (heat), or solvent means. Nonwoven fabric production usually includes four stages: (1) fiber preparation, (2) web formation, (3) web bonding, and (4) post treatment. Nonwoven fabrics constitute one of the fastest growing segments of the textile industry. Examples include nonwoven interfacings such as Pellon and felt.

DYEING

Dyeing can be done at any stage of fiber, yarn, or fabric production. Some of the most important methods of dyeing are the following :

Producer colored. Producer or solution dyeing, used for man-made fibers, adds the pigment or color when the fibers are still in solution; before the filaments are formed.

Stock dyeing. The dyeing of loose fibers before yarn processing.

Yarn dyeing. The quality method of dyeing certain woven patterns such as stripes, plaids, and checks. It is done after the yarn is spun but before weaving or knitting. It is most often used in shirts.

Piece dyeing. The dyeing of a piece of fabric after weaving or knitting. It is the least expensive, most widely used method of dyeing solid colors.

Piece dying.
(Courtesy of the International Wool Secretariat, England)

Cross dyeing. A type of piece dyeing that achieves a simple, less expensive two-color pattern. The cloth must be made of fibers having affinities with different dyes, so that the cross dyeing can achieve varied color effects in one dyebath.

Garment dyeing. Greige goods are preshrunk and sewn into garments. The whole garments are dyed after they have been sewn. This method insures quick delivery of needed colors as well as uppers and lowers that match perfectly.

Computers have been enlisted to help regulate dye mixing to try to have colors match from lot to lot.

If apparel manufacturers want a fabric in a color that is not regularly available, their order must meet a minimum yardage requirement. Yarn-dyed and solution-dyed fabrics require the largest orders. This has forced many manufacturers to seek fabrics outside the United States where minimums are lower. In turn, domestic producers are trying to be more flexible.

PRINTING

Printing is used to apply design or pattern to fabrics. There are two basic printing techniques: wet printing and dry printing.

Wet Printing

In both engraved-roller printing and screen printing, dyestuffs are applied wet for optimum color penetration. In pigment printing, one method of wet printing, the pigment is attached to the fabric surface with a resin. Other wet prints use dyestuffs that have a chemical affinity to the cloth fiber and do not require resin. With wet printing it is possible to achieve a soft, drapable hand as well as a crisp finish.

Engraved-Roller Printing

In this technique, a separate roller engraving is used for each color in the pattern. The design is rolled onto the fabric as it passes through the printing machine.

Screen Printing

Flatbed screen printing uses a screen spread over a frame. The portions of the design to be printed are made of porous nylon fabric that allows the color to pass through the screen. The areas that are not to be printed are covered or coated with enamel. Color is poured into the frame shell and is forced through the nylon by means of a squeegee worked back and forth. Flat-bed screen printing is versatile but expensive. In Como, Italy, artisans use as many as 50 silk screens with separate colors (stencils made of wire mesh) to print one scarf in perfect registration.

Rotary screen printing is a mechanized version of flatbed screen printing. In this method, the roller itself is porous in the areas to be printed. Dye is forced into the roller cylinder and through its porous screen as it rolls over the cloth. This method is much faster than flatbed screen printing and is continuous, leaving no breaks between screens.

Flat bed screen printing.
(Courtesy of the International Wool Secretariat, England)

Dry Printing

Heat-Transfer or Paper Printing

In this process, rotary screens or rollers first print dyestuffs onto paper. The paper can be kept for use at any time. This process helps keep converters' fabric inventories lean since greige goods are printed against orders and not for stockpiling.

To print on fabric, the paper and fabric are put through hot rollers; the dyestuffs sublimate into a gas, which moves from the paper base onto the fabric. The advantages of this method are that it gives a clean, fine line on knits with no mistakes or waste and paper is a smaller investment than the elaborate equipment needed for the other methods. However, this method can cause the hand of the goods to become stiff and the printed pattern transfers only to the fabric surface, with little penetration, creating potential *grin-through* (fabric showing through) problems.

The demand for prints, as well as developments in low-sublimation dyes, deeper-penetrating dyes, and better inks for a variety of fiber types, has given dry printing wider acceptance. There is relatively little waste water or harmful discharges with this method.

FINISHING

Finishing is the term used to encompass all the processes used to enhance a fabric (usually after dyeing or printing). Finishes can radically alter fiber and fabric characteristics, performance, or hand. Here are a few of the more common finishes:

Calendering. A mechanical process of passing fabric between heavy rollers. By using different combinations of heat, pressure, and rollers, it is possible to produce a wide assortment of effects such as glaze, watermark, or moiré. Calendering is usually done on synthetics, because it is not permanent on fabrics made of natural fibers.

Caustic reduction. A process usually done on polyester to give it a silklike feel. The surface of the fibers are eaten away in a caustic bath, which reduces the weight of the fabric.

Durable press. The application of certain resins to cotton to create fabrics that require little or no ironing (also called *permanent press,* although it is rarely permanent).

Heat setting. A final finishing that heats thermoplastic man-made fabrics to just below their melting point. This treatment stabilizes them so there will be no further change in size or shape. This process can be used to permanently set new forms such as pleats.

Mercerizing. The treatment of cotton with a cold, strongly caustic chemical solution to achieve a lustrous silklike finish.

Sanding (or sueding). The process of mechanically rubbing the fabric with rolls coated with fine grit sandpaper to create a soft surface.

Napping or shearing. Fabric surfaces are raised and plucked with needles on rotating drums. These fabrics are usually sheared to give the surface a uniform pile or flannel surface.

Shrink control. The preshrinking of cotton cloth so that it will not shrink during laundering.

Water repellency. Yarns treated with chemicals, then woven to create cloth that permits air and vapor to pass through the fabric while keeping rain and snow out.

Other finishes can make fabrics flame retardant, fade resistant, mildew resistant, bacteria resistant, or stain resistant. See Chapter 10 for information on garment finishing.

When production is complete, fabrics are measured and rolled onto tubes. Each finished roll, called a *piece,* may have 40 to 100 yards on it, depending on the weight of the goods (heavier knits and wools have less yardage on a piece to make them easier to handle). The rolls are shipped to manufacturers to fill production orders for multiple pieces that can amount to thousands of yards. Smaller quantities, in the form of 3-yard to 10-yard sample cuts, may be ordered for designer styling, or duplicate yardage (perhaps 100 yards) can be ordered for sales representatives' samples.

SUMMARY

Textile producers, the suppliers to apparel manufacturers, are an integral part of the fashion industry. Textile fibers are the basis for fabrics. They are classified as natural (cotton, flax, wool, or silk) or man-made (cellulosic or noncellulosic). Yarns are made from fibers or filaments and are usually woven or knitted into fabrics. The fabrics are then dyed or printed and finished in preparation for shipment to manufacturers.

The phenomenal growth of the textile industry would not have been possible without the development of man-made fibers. These fibers have revolutionized the industry, especially in the areas of yarn and knit production and fabric printing. The industry is making advances in the development of environmentally friendly production. The textile industry is now dominated by a few large companies for only they have the capital for the modern technology needed to survive in today's competitive international textile market (see Chapter 6).

CHAPTER REVIEW

Terms and Concepts

Briefly identify and discuss the following terms and concepts:

1. Textiles
2. Natural fibers
3. Flax
4. Ramie
5. Worsted yarns
6. Synthetic fibers
7. Man-made fibers
8. Cellulosic fibers
9. Lyocell
10. Microfibers
11. Generic fiber names
12. Man-made fiber spinning methods
13. Spun yarns
14. Filament-yarn texturing
15. Plain weave
16. Twill weave
17. Warp knitting
18. Types of dyeing
19. Printing techniques
20. Finishing methods
21. Mills
22. Converters

Questions for Review

1. Why is a study of textiles important to someone working in apparel production?
2. What determines the quality of cotton?
3. What are the two basic ways of making fabric?
4. How is knit yardage made?
5. Give examples of textile dyeing methods that can be used at three different levels of fabric production.
6. Why were specialization and consolidation important in the textile industry?
7. What is the difference between a mill and a converter?

Projects for Additional Learning

1. Examine the fiber-content labels in the clothes in your wardrobe. How many garments are made from natural fibers? How many from man-made fibers? How many from blends? How does the fiber content affect the care of the garment?
2. Visit a local fabric store and find examples of a plain weave, a twill weave, a satin weave, a novelty weave, a warp knit, a circular knit, and a nonwoven fabric. Ask for tiny swatches of each or purchase the smallest amount possible to illustrate your findings.

NOTES

[1] Interview, April 28, 1992.

[2] As quoted in "Textile Report," *Women's Wear Daily*, January 14, 1992, p. 12.

[3] "World Fibre Production," *Wool Facts* booklet, International Wool Secretariat, London, March 1992, chart 1.

[4] "Imports," *Women's Wear Daily*, January 28, 1992, p. 27.

[5] Berrye Worsham, Cotton Inc., Interview, February, 1993.

[6] *Wool Facts*, Tables 52 and 2.

[7] *Wool Facts*, Table 2.

[8] *Wool Facts*, Tables 23 and 24.

Joyce Perkins, Assistant Manager, Apparel Fashion Marketing at Cotton Incorporated, making a presentation in Thailand. *(Courtesy of Cotton Inc.)*

TEXTILE PRODUCT DEVELOPMENT AND MARKETING

CAREER FOCUS

In textile product development there are career opportunities in research, styling, and merchandising. In marketing, positions range from entry-level junior sales representative to account manager, sales manager, marketing manager, and director of marketing. There are also the support areas of advertising and public relations to consider. Most of these opportunities are in New York City, Los Angeles, or other textile centers.

It is also essential for manufacturers and retailers to understand the marketing forces that transform fibers into fabrics and then move those fabrics into the hands of the apparel designers and manufacturers.

CHAPTER OBJECTIVES

After reading this chapter you should have attained competence in the following areas:
1. Understanding of the impact of imports upon the domestic textile industry
2. Understanding of the role of Quick Response in the textile industry
3. Understanding of the importance of product development for both fibers and fabrics
4. Knowledge of the promotional tools of the textile industry

*T*extile product development and marketing are essentail activities through which the textile industry responds to consumer demand. The consumers include the fabric manufacturers who use the fibers, the apparel manufacturers who use the fabrics, and the end user who buys the apparel. *Product development* covers the research, merchandising, and styling involved in creating new and updating existing products. *Marketing* incorporates the entire process of planning, promoting, and selling of goods, in this case textiles.

Much has changed and will continue to change in textile marketing. This chapter discusses the impact that foreign competition and consumer demand has on the industry. It also describes how the industry is reacting with product development, new technologies, and new marketing strategies for fibers and fabrics.

THE GLOBAL TEXTILE MARKET

We can see globalization in every aspect of our lives. To learn about fashion, we need to know how globalization affects the textile industry.

One of the major trends in the textile and fashion industries is globalization. Nowhere is this more evident than in the textile industry. Ever increasing amounts of textile products are being imported into the United States because of the availability of cheaper labor abroad. U.S. manufacturers compete with labor rates of as little as 25 cents per hour in China, Thailand, Pakistan, and India and 15 cents an hour in Indonesia.[1]

Imports Cause Trade Imbalance and Loss of Jobs

Fierce competition from imports has caused the American industry to lose over 50 percent of its domestic market and has resulted in buy outs, takeovers, and consolidations to salvage what was left. Since 1980 imports have increased from 4 billion square meters to 12.4 billion; the textile trade deficit has increased from $5 billion to $25 billion; the industry has lost at least 350,000 jobs; and 1,000 plants have closed.[2] This corresponds to the enormous growth of the textile industries of Japan, China, Taiwan, Korea, and India. In addition, since 1992, when Europe became an integrated economic community with no trade barriers between countries, it has a textile industry larger than that of the United States. The Eastern European countries are a huge potential competitor as well.

Textile industry leaders realize that imports will continue to grow, but they want some restrictions on them so that the domestic textile industry can survive. The U.S. government, however, argues that tariffs on textiles and apparel at an average of 18 percent are already three times the rate of those on other manufactured products.

General Agreement on Tariffs and Trade (GATT)

The objective of GATT, a contract between 103 governments, is to provide a secure international trading environment and a continuing process of trade liberalization. It sets multilaterally agreed rules governing trade behavior, acts as an international court to resolve disputes between members, and serves as a forum for trade negotiations. Representatives for GATT members have met at eight negotiating conferences known as *rounds*. In the Uruguay Round begun in 1986, GATT members continue their talks to set broad general codes of conduct and to decide the fate of the Multifiber Arrangement.

Multifiber Arrangement (MFA)

In 1974, international trade in textiles and apparel of cotton, wool, and man-made fibers came under the jurisdiction of the Multifiber Arrangement. The MFA sets up rules regarding exports, imports, and quotas. The MFA is scheduled to be phased out within ten to fifteen years. However, before that happens, the U.S. would like to improve access to other nations' markets and get tariff concessions from trading partners. It is expected that import duty will remain at an average 16 percent on most textiles and apparel even at the end of the phaseout period.

Quotas on Imports to Protect Domestic Industry

The United States government has set quotas on imports to try to protect the American textile industry. However, there are no quotas on silk, flax, and ramie because they are not produced in the United States and therefore are not considered to compete with American industry. The U.S. also faces a problem caused by *transshipping* (shipping merchandise from the manufacturing country to a country with quota available, before shipping to the United States) to circumvent apparel quotas. One bright spot for domestic textile producers is the shift towards apparel production in the Caribbean Basin to take advantage of new government programs that permit quota and duty breaks if domestic fabrics are used (also see Chapter 10).

North American Free Trade Agreement (NAFTA)

A proposed trade agreement, NAFTA, would create a free market (devoid of import duties) of 560 million people in Canada, the United States, and Mexico. Canada and the United States already have a free trade agreement that went into effect in 1989 and Canada remains the leading market for U. S. exports of textile mill products.

There is controversy as to whether NAFTA will help or hinder the American textile industry. Supporters feel that NAFTA will give textile firms the opportunity to ship fabric to Mexico to be made into garments and curb the steady increases of production in the Far East where Asian fabrics are used. M. L. Cates, Jr., president of the American Textile Manufacturers Institute and Arkwright Mills, said, "We think NAFTA will give our industry a much needed shot in the arm, enabling us to work with Mexico and Canada for mutual new growth opportunities."[3] The domestic textile industry wants to give the Caribbean Basin countries the same preferred treatment as Mexico. Opposers feel that more jobs will be lost to Mexico as we cannot compete with their low wages.

QUICK RESPONSE

Computer technology has helped textile producers keep track of the flow of goods from fiber to fabric to apparel, cut down on wasted time, and speed reorders.

In an effort to combat imports with faster response to customer needs than imports can provide, textile and computer executives have developed a computer strategy called *Quick Response (QR)*. This is an attempt to reduce waiting time in ordering and distribution between textile and apparel producers (and retailers) by forming cooperative alliances between all levels of the industry via *EDI* or *Electronic Data Interchange* (the exchange of business data between two parties by means of computer). Supporters of Quick Response point out that the greatest dollar loss to industry is time lost in distribution from fiber to final sale.

This strategy requires a huge financial outlay for technology that is ideally recuperated in larger sales. All goods are given *universal product codes (UPC)* which identify style, color, size, price, fabrication, and vendor. These bar codes are laser scanned and the information is fed through the EDI pipeline to manufacturers. When a particular style is selling well at retail, coded information on that style (including the best-selling colors, fabrics, and sizes) immediately informs the retail buyer to replenish stock, the apparel manufacturer to issue new cuts, and the fabric producer to send more fabric to the apparel manufacturer. Of course for this strategy to work, it needs the complete cooperation of each level of the industry, a willingness to supply on a reorder basis, and the flexibility to sell smaller quantities of fabric and issue smaller, more frequent cuts. Linkage systems from one branch of the industry to another and standardized codes for data exchange throughout the industry have been established.

The goal of the Quick Response strategy is to enable the fabric producer to supply and resupply the manufacturer upon shorter notice. Issuing smaller initial cuts until a style is tested will prevent markdowns, and quickly replenishing stock with best sellers will mean higher sales. This allows product development to be done closer to the selling season in order to better anticipate consumer preferences.

Although electronic data interchange has become important to the industry, many converters and apparel designers feel that QR is suited primarily to basic goods and volume manufacturers. Although QR has been successful for volume manufacturers, supporters still have to prove its value on the designer level. In any case, computers have become a necessary part of textile marketing. Without them it would be nearly impossible to keep control of volume production and inventory in today's overwhelming megamarket.

In conjunction with Quick Response, textile producers formed the Crafted with Pride in the USA council spearheaded by Roger Milliken (chairman of Milliken & Co.) to promote domestic-made goods. The council points out that the American textile industry has the advantage of its close proximity to the American consumer. In theory, domestic suppliers should be more convenient, dependable, and expeditious. Purchasing U.S.-made textiles should mean that the buyer has better control and doesn't have to commit to an order so early. This would help American manufacturers and retailers who want to give their customers what they want in increasingly shorter lead times.

However, many people now feel that Crafted with Pride is outdated—that value, cost, style, color, fabric, fit, and fashion are the real factors that influence consumer purchasing, not the country of origin. W. Matt Self, president and CEO of Greenwood Mills, commented, "...the consumer is more attentive to quality products at reasonable prices than to patriotic approaches."[4]

NEW MARKETING STRATEGIES

To compete with imports, textile producers have developed new marketing strategies.

Many apparel manufacturers and retailers argue that textile leaders should spend less time lobbying for import duties and quotas and more time on research and development in order to have more innovative styling, more flexibility, and better quality as a preferable means to fight imports. Many apparel and retail executives feel that American textile firms are not as innovative as Japanese and Italian companies.

In response to consumer demand, textile producers have had to overhaul their marketing strategies completely. With so many diverse life styles and resulting market niches, textile firms realize that they have to be *market-driven* (responding to market or consumer needs) rather than manufacturing-driven.

Because the modern consumer wants uniqueness and variety, textile manufacturers must not only respond more quickly to the market but must be flexible and provide innovative styling. To be fashionable and flexible, it is necessary to produce smaller quantities; however, producing less costs more per yard. It has been the American way to produce large volume so when the market changes it is difficult and expensive to make changes. Americans streamline the product in order to be efficient and to offer quality at a reasonable price. The new strategies are more difficult; it is easier and less expensive to go on making the same thing. However, innovation, which requires creativity and not just improved technology, has become an important part of the business.

FIBER PRODUCT DEVELOPMENT AND MARKETING

In order to meet consumer demand, fiber producers do extensive research and product development. Textile marketers then promote and sell these fibers to yarn or fabric producers.

In the past, the marketing role of the natural-fiber producer was relatively simple. The crops or animals were raised, harvested, or sheared, and the fibers sold at local markets to wholesalers who in turn sold the fibers at central markets. Sheep farmers and cotton growers did not have to be concerned with fabric and garment production.

The development of man-made fibers and new marketing strategies changed all that. Now producers try to create new variants or find new uses for existing natural fibers to meet demand. Each company tries to fill market needs but in a different way from the competition. Consumer or industry demand for certain fiber or fabric properties, however, may cause several textile companies to research the same problem simultaneously. As a result, several companies may develop similar, competitive fibers.

Fiber companies invest a great deal of time and money in research and development. Several years may be required to develop new fibers or variants (see Table 6–1). Natural fiber producers, because of their generally small size, usually conduct their research together. To do this, they have formed associations representing the producers of a particular fiber. Among them are the International Institute for Cotton (Cotton Incorporated in the United States); the National Cotton Council of America, the International Wool Secretariat (the Wool Bureau); the Silk Institute; and the International Linen Promotion.

Sales Promotion

To foster sales, fiber producers must make their products known to potential consumers. The man-made fiber industry has concentrated on consumer brand recognition to explain what a new fiber or variant is and what uses it has. Natural fiber associations also want the public to be

TABLE 6-1
Timing of Product Development in the
Textile and Apparel Industries
(The industry constantly strives to shorten lead times.)

Activity	Length of Time Before Retail Selling Season
Development of new fibers and variants	Several years
Fabric development (fiber firms working with mills to research and develop new fabrics)	1 to 2 years
Color predictions (fiber level)	18 to 20 months
Presentation of new fabric lines by fabric producers (Interstoff, Premier Vision, Ideacomo, etc.)	1 year
Shopping fabric lines (by designers and merchandisers)	8 months to 1 year
Apparel design and line development	6 to 9 months
Apparel collection openings, market weeks, showing lines to retail buyers, taking orders	4 to 6 months
Apparel production	1 to 5 months
Shipping to retail stores	1 week to 1 month
Apparel for sale in retail store	0

aware of the attributes of their fibers. Both use trademarks or logos for identification, such as the Wool Bureau's ball of yarn symbol.

To prepare for international textile trade shows such as Premier Vision or Interstoff, fiber producers may have sample fabrics and garments made up to display. These samples show yarn and fabric producers how the fiber may effectively be used and how they may promote a sale.

Advertising and publicity

Fiber producers advertise to fabric and apparel manufacturers in trade publications. To reach the general consumer, they offer *cooperative advertising dollars* (to share the cost) to manufacturers and retailers. These cooperative allowances are based on percentage of sales of fibers used in fabrics sold to the manufacturer or in garments sold to the retailer. Some allowances are based on market potential in order to stimulate a new market or stimulate interest in a fiber. In addition, fiber firms often provide ad materials, such as photographs.

Publicity is another way of sending a message to the consumer. In this case there is no cost for media coverage. Fiber producers and trade associations continually provide the press with newsworthy material in hope of an editorial mention.

Marketing and customer services

Fiber producers provide many services to their customers on both the manufacturing and retailing levels in order to promote sales:

Consumer education. Fiber producers continually provide information on their products to apparel manufacturers, schools, and the general public in the form of brochures, exhibits, lectures, audiovisual aids, and films. This information covers every aspect of fibers and fabrics including history, production, use, care, and benefits to the consumer.

Technical advice. Fiber producers offer advice to yarn and textile mills as well as to garment manufacturers. Fiber producers often commission fabric designers to work up new ideas for them. Samples of these ideas are woven or knitted in short runs, and swatches are sent to the mills. The fiber producers are also able to offer mills production advice.

Hang tags. Some fiber producers provide apparel manufacturers with printed hang tags listing care instructions. These can be ordered by the garment manufacturers through the fabric producer. Hang tags can be a useful form of publicity as well as information for the consumer.

Fashion presentations. Fiber producers have been the major source for styling and color directions. Based on their research, fiber producers make trend direction information available to mills, converters, and manufacturers, usually on a semiannual basis. A stylist representing the fiber producer may make presentations to manufacturers showing garments, new and experimental fabrics, and color charts. Presentations are often targeted to specific needs such as coat ideas for coat manufacturers or men's wear ideas for men's wear manufacturers. Market research and sourcing information may be included. Some companies also produce fashion styling reports.

Color forecasts. Color directions based on color cycles, the economy, the arts, and international fashion trends are researched at the fiber level. These are combined with basic and seasonal colors to round out the

A selection of wool color cards and yarn forecasts.
(Courtesy of the International Wool Secretariat, England)

color story. Companies decide on the colors they think will be the most popular and dye yarn or fabric swatches to present to their customers. These are usually made available to fabric producers and manufacturers to help them plan their color stories.

Fabric libraries. An important service of the fiber producers is their fabric libraries, located in major fashion centers such as Paris, London, New York, and Los Angeles. Fabric libraries keep samples of fabrics from every mill or converter using their fibers. A designer can visit the library for an overall picture of what is available. If a designer is looking for a specific fabric, the library can help locate a mill or converter that makes it. Some libraries now bar-code the fabric samples. A scanner automatically reads and prints out the name and address of the mill and the contact person. At the same time, the designer's name is sent to the mill or converter so they know of potential customers.

Fiber Distribution

Natural fibers

Fibers are sold to mills for yarn spinning and weaving or knitting. The farmers who produce natural fibers sell their goods at the markets organized by their various trade associations. The major markets in the United States are Dallas, Houston, Memphis, and New Orleans for cotton, and Boston for wool. Farmers have no control over prices, which are set in the marketplace by supply and demand.

Man-made fibers

The large chemical companies that produce man-made fibers have their own sales forces and set prices based on their costs. In some cases in the Far East, when fiber producers and mills are vertically integrated into one large company, the producing company becomes its own market.

Fibers are often sold under certain obligations so that standards of quality may be controlled. After spending a great deal on advertising to build a good reputation for a fiber's performance, the producer wants to preserve that good name. Restrictions on standards are therefore imposed when the fiber is sold under a brand name or a licensing agreement. There are no restrictions if the fiber is sold without the use of a brand name.

FABRIC PRODUCT DEVELOPMENT AND MARKETING

The next step in the textile marketing chain is the development and marketing of fabrics.

Fabric is becoming the driving force of change within the fashion industry. Karl Lagerfeld commented, "Fabrics bring the big changes now. Lycra, stretch and all that really made things possible that you could not do before, except in couture with 20 fittings."[5] There are many exciting innovations in fabrics especially in blends which create interesting new textures.

Fabric producers try to anticipate consumers' needs by doing intensive market research; this helps them compete for the business of the apparel manufacturers, their direct customers. The textile industry must develop fabrics early enough to allow designers time to find and test suitable goods. Fabric producers also study international fashion and fabric trends and work to develop fashionable products having desirable fiber blends, finishes, and other properties. In turn, they offer their expertise to apparel manufacturers and sell *sample cuts* of fabrics for test garments.

Textile Design

Textile designers create print designs for fabrics or suggest styling ideas for a weave or knit. They work hand in hand with textile engineers and have to understand textile processes to know if their ideas will work. Their ideas come from worldwide influences (see Chapter 4), but especially from historic and folkloric motifs, wallpapers, old fabrics, and nature. Many fabric companies maintain fabric archives to supply designers with ideas from the past for new prints.

Textile designers must consider the essential elements of color, texture, line, shape, and space. They are concerned primarily with a two-dimensional surface, the flat fabric, rather than with the three-dimensional human form. Yet they must keep the end use of the fabric in mind if they are to create practical designs. Textile designers may use CAD systems to help them experiment with visual representations of weave and pattern designs when designing a collection of fabrics to present to their customers. There must be a variety of fashionable and basic prints, solids, and textures in a collection. Designers at Abraham AG, a trendsetting converter in Zurich, for example, create collections of more than 300 pieces twice a year.

A textile designer uses a computer-aided-design system to experiment with color combinations in a print repeat. *(Courtesy of Cotton Inc.)*

Designer Marc Grant paints one of his original textile designs.
(Courtesy of Jeanne-Marc, photographed by the author)

Print design

Representatives from mills or converters may shop worldwide for fabrics or garments as a source of new ideas. Tradition has centered print studios in England, France, and Italy. Mills, converters, and sometimes large garment manufacturers visit studios seasonally to puchase original *croquis* (paintings) which are then put into a repeat design for fabric application. Larger print studios bring their design collections to New York, Tokyo, Hong Kong, and other textile centers for trade exposure. Variations, repeat work, restyling, and coloring may be done by the converter's studio or by an independent design studio.

The motifs for a print should have interesting shapes, a pleasing rhythmic pattern, and a harmonious relationship to one another. No single part of the overall pattern should claim too much attention. Print design is a continuous repetition of motifs, and the designer must consider how cutting will affect the pattern. Different types of prints may be designed for particular segments of the apparel industry. For example, florals and feminine motifs may be aimed at women's apparel, traditional geometrics at men's wear, and small, whimsical prints at children's wear. However, the use of prints is also cyclical; some kinds of prints are more popular than others at a given time.

Coloring

A print is usually offered in several colorways. For a new color combination, a croquis is made by the textile firm's design department or by computer. If that is approved, a *strike-off* is run as a test on a short piece of fabric. Minimum yardage requirements for custom colors vary with the printing method.

Video

Some people feel that video may be the communications link of the future between levels of the industry. Video proponents point out that since consumer decisions are made visually, industry decisions should be made likewise. For example, a fabric designer and an apparel designer could have an interactive video dialogue about developing a suitable fabric for the designer's needs. This method would allow apparel designers to express their wants and needs as the fabric is being developed.

Sales Promotion

Fabric producers compete for the business of apparel manufacturers and for retail-level acceptance of products made of their goods. They use advertising, publicity and customer service to promote their products.

Advertising

Large fabric producers advertise the brand names of their products. Like the fiber producers—and often in cooperation with them—they use

A cooperative advertisement between Burlington, the fabric producer, and the Wool Bureau. Note: wool symbol at lower right. *(Courtesy of Burlington Industries, Inc.)*

television, newspapers, and magazines to reach national audiences. They also join apparel manufacturers and retail stores in cooperative advertising nationally or locally. Co-op ads must contain the brand names or trademarks of all participants .

Publicity

Many fabric producers also provide educational materials to consumer groups, schools and colleges, and the general public. They also supply information on fabric fashions and developments to the press in hope of editorial coverage.

Customer service

Some fabric producers also employ merchandising and marketing experts who analyze trends and pass the information on to manufacturers and retailers. These experts provide printed or visual materials to manufacturers and present in-store fashion shows and sales training sessions for retailers.

Fabric Markets

Twice a year, textile producers from all over the world display their lines at important international yarn and fabric shows.

Pitti Filati ("Pitti Yarns," originally shown at the Pitti Palace). Held in July and January in Florence, Italy, this is the foremost international yarn show.

Yarn Fair International. This is New York's yarn show held each August and February primarily for Americans who are unable to go to Florence.

Premier Vision ("First Look"). With approximately 650 exhibitors and 40,000 visitors, this market held in March and October in Paris, has become increasingly important as an early source for fashion fabrics.

International Fashion Fabric Exhibition (IFFE). The newest New York show of fabrics in March and October; it is scheduled between Premier Vision and Interstoff to give European textile producers a chance to do business in the U.S.

Interstoff ("International Fabrics"). Held in April and October or November for 3 days in Frankfurt, Germany, Interstoff offers buyers and designers an opportunity to see the complete lines of over 1000 exhibitors.

Ideacomo ("Ideas from Como"). Primarily an exhibit of Italian silks, this market follows Interstoff in April and November in Como, Italy.

The entrance to the Premier Vision fabric fair in Paris.
(Courtesy of Premier Vision, Lyon, France)

There are also fabric shows in other major cities. Textile trade organizers often try new shows such as Interstoff Asia to try to make buying convenient. Textile producers visit these shows to see the latest directions in fabrics. Manufacturers and their designers visit the shows to buy fabrics and gather ideas about how to use them.

There are two official fabric seasons per year, however, the market is becoming increasingly seasonless and new fabrics are continually being developed.

The Sales Representative

In addition to attending the textile shows, fabric producers maintain showrooms in major fashion centers to display their new lines of fabrics. Each fabric company also has sales representatives who periodically visit manufacturers and their designers, showing them suitcases full of *headers* (head ends of fabrics).

Most salespeople in the textile industry, especially for large companies, are salaried employees of the firms they represent. Independent sales agents or representatives may handle several textile lines and are paid by commission (usually 3 percent of the price of fabrics sold and shipped). Salespeople are assigned specific manufacturers to call on. They try to make contacts to gain new accounts, as well as to show and sell fabrics to regular customers.

The Jobber

Another independent agent who deals in fabrics is the jobber. *Jobbers* purchase fabrics in quantity from manufacturers and resell them. They may buy from several sources and put together their own line, or they may buy goods in hopes of selling them later at a good markup during a period of peak demand. Buying leftover fabrics clears warehouse space, gives operating cash to the textile producer, and makes these fabrics available to a manufacturer late in the season. Another role of jobbers is to buy up unsold goods at great reductions, for resale to discount stores or other outlets. As compensation for the financial risks jobbers take, they can buy fabrics at considerable discounts, which enables them to make a profit on resale.

SUMMARY

The tremendous growth of the Asian textile industry and its low production costs due to cheap labor have caused keen competition for the U.S. textile industry. The United States is fighting imports with new marketing strategies to meet consumer demand and new technology such as Quick Response to speed distribution in the marketing chain. Product development has become very important and the industry is striving for innovation and flexibility. The industry promotes fibers and fabrics with brand-recognition advertising and customer services. International fabric markets and sales representatives provide the link to the apparel manufacturer, the next level of the fashion industry.

CHAPTER REVIEW

Terms and Concepts

Briefly identify and discuss the following terms and concepts:

1. Quotas
2. GATT
3. MFA
4. Quick Response
5. Marketing strategies
6. Market driven
7. NAFTA
8. Brand recognition
9. Cooperative advertising
10. Customer services
11. CAD
12. Strike-off
13. Premier Vision
14. Headers
15. Jobbers

Questions for Review

1. What is color forecasting?
2. Describe the services of a fabric library.
3. How does modern marketing differ from marketing in the past, when only natural fabrics were available?
4. How does modern technology help to fight imports?

Projects for Additional Learning

1. Select a print that you like from a book on historical costume or art. Trace (or modify to update, if necessary) and recolor the print in the current season's fashion colors.
2. Shop a local department store to compare fabrics used in designer or contemporary fashions.
   a. What fabrics look the freshest and most exciting?
   b. What kinds of fabrics (knits, wovens, prints, solids) are used the most?
   c. What are the predominant colors?
   d. Examine hang tags and labels for fiber content. What kind of fiber is used more frequently in the garments—natural or man-made?
   e. Can you identify the fiber brand names and generic names on the labels?
   f. Can you identify fabric producers by information on the labels?
   g. Is enough information provided for the consumer about the fibers and fabrics?
   h. Summarize your findings in a written report.
3. Shop a local department or specialty store and examine country-of-origin and fiber-content labels in twenty garments.
   a. Among the garments made of cotton, what country of fabric origin predominates? Where is each garment made?
   b. Find the same information for garments made of wool, silk, ramie, and man-made fibers.
   c. Make a chart showing fiber content, country of fabric origin, and country where made. Make separate sections for (1) foreign fabric and labor, (2) imported fabric but American labor, and (3) American fabric and labor.
   d. Do you see a pattern forming? What conclusions can you draw from these data?
   e. Are all garments adequately labeled?
   f. How do prices compare between garments made overseas and those made in the United States? How do quality and detailing compare?

NOTES

[1] "Textile Report," *Women's Wear Daily*, January 28, 1992, p. 27.

[2] Ibid.

[3] As quoted in "ATMI Calls '92 a Better Year," *Women's Wear Daily*, December 30, 1992, p. 8.

[4] As quoted in "Is it Time for a Change?" *Women's Wear Daily*, December 2, 1991, p. 4.

[5] As quoted in "King Karl," *Women's Wear Daily*, November 20, 1991, page 8.

Lace trimmings are important to the Jessica McClintock look. *(Courtesy of Jessica McClintock)*

7

TRIMMINGS, LEATHER, AND FUR

CAREER FOCUS

There are many job possibilities in the trimmings area, very similar to the textile industry: stylists, merchandisers, technicians, and marketers. Much of the domestic leather and fur industry has been lost to imports but there are opportunities in marketing.

Designers and manufacturers who use trimmings have to learn about them to use them properly. Apparel manufacturers need trimmings buyers. Retailers need to know the sources and characteristics of the fur and leather they sell.

CHAPTER OBJECTIVES

After reading this chapter you should have attained competence in the following areas:
1. Knowledge of how threads, interfacings, narrow fabrics, zippers, buttons, and belts are produced and used.
2. Understanding of the sources of fur and leather and the steps in processing them.

To complete our discussion of the raw materials of fashion, we are including trimmings, leather, and fur.

Trimmings, covered in the first half of this chapter, are necessary to complete a garment. The category of trimmings is very diversified, comprising both textile and nontextile areas. Thread, interfacing, and narrow-fabric manufacturing are extensions of the textile industry having similar production and marketing procedures. Zipper and button manufacturing are entirely separate industries with their own resources and production methods.

The second part of the chapter is devoted to the sources and treatment of leather and fur. Leather and fur were used for clothing long before textiles were developed. Even today, leather and fur are important materials for the fashion industry. Leather's essential use is in fashion accessories—shoes, handbags, gloves, and belts—whereas fur is primarily a garment material.

Although much older than the production of fiber and fabrics, the processing of leather and fur is not as sophisticated—and it takes much longer. Lately, however, production methods have improved, greatly increasing the supply and variety of leather and fur. However, as with textiles, the domestic industry has shrunk considerably due to competition from imports.

TRIMMINGS

Designers and trimmings buyers need to know the best uses of a wide array of trimmings.

Trimmings are the necessary supplies used to finish and adorn both garments and accessories. Decorative trimmings include buttons, buckles, belts, braids, ribbons, fringes, bows, laces, and much more. Functional trimmings, such as elastics, tapes, interfacings, threads, zippers, and shoulder pads are sometimes referred to as *findings*.

The Trimmings Expo is held each November in New York for designers and manufacturers to see the newest in this important market. Trimmings buyers for apparel manufacturers must find the best trimmings available for each kind of application in each garment, make sure that the colors of all trimmings used in each garment match, and coordinate deliveries from each trim producer so that all trims are in house in time for manufacturing. The trimmings buyer at Koret, for example, must purchase an average of five different trimmings and findings per single fabric order—or 3000 trims compared with only 600 fabrics ordered per year.

Threads

Thread is supplied by yarn producers. Formed by spinning and twisting textile fibers or filaments together into a continuous strand, it is vital in determining the quality of workmanship in a garment.

Until recent years, cotton and other natural fiber threads met requirements of durability, appearance, and sewability because the majority of fabrics available were also made of natural fibers. However, the advent of man-made fibers brought about the development of many new fabrics, including knits. Because of their new characteristics, these fabrics demanded a stronger, more elastic thread. To meet these specifications, polyester threads were developed.

Thread choices available today include:

Cotton, used in cotton and wool garments.

Silk, used in silk garments and men's wear.

Nylon, a monofilament used primarily in men's wear.

Rayon, or rayon wrapped around a nylon core.

Polyester and polyester blends:
  A polyester core wrapped in cotton (to give flexibility to the cotton).
  Three-ply 100 percent spun polyester.
  Long-staple polyester.

Threads are wound on cones in 1200-yard to 24,000-yard lengths for factory use. The Thread Institute has suggested metric conversions that approximate current yard lengths. For example, the 6000-yard cones will be converted to 5500 meters (actually 6010 yards).

Manufacturers use various thread sizes—that is, various thicknesses of thread—according to fabric and quality needs. A metric system called *tex* has been suggested for measuring thread size. Based on the weight of the raw material, the tex number increases as the size increases; average thread sizes range from 10 to 500.

Elastics

Elastic is used more often in today's clothes, which include so many pull-on pants and skirts and stretch cuffs. Of course it has long been used in underwear and swimwear.

Elastic is primarily a man-made rubber called *elastimer.* It can be woven, braided, knitted, cut into flat strands, or made into thread. The woven type (including webbed) is regarded as the best for use in multi-stitch or tunnel waistbands because the elastimer strands are covered (wrapped) with polyester or rayon and can better control stretch and recovery, important considerations for a manufacturer. Monofilament versions such as XL90 (which looks like a screen when woven) have been developed to help prevent rolling. Webbed elastic wrapped in colored thread is used for decorative purposes such as stretch belts.

Less expensive braided or knitted elastics that narrow when stretched can be used in narrow widths to gather the wrist of a blouse, or for other purposes where roll is unimportant. Flat elastic is usually reserved for underwear. Because it is a solid band, flat elastic breaks when a needle goes through it; with the other varieties, the needle goes between the strands. Elastics are available in widths from 1/8 inch to 3 inches to fit manufacturers' needs.

Elastic thread (a strand of elastimer wrapped in cotton, rayon, or polyester) is sewed into fabric to add stretch, as in shirring or smocking. It can also be used as loops for buttons. Of course there are other elastic fibers such as spandex, which are used to make the stretch fabrics so important to the swimwear and active sportswear industries.

Interfacings

Reinforcement is needed to give a garment support and to keep it in shape longer. The most essential kind of inner construction is the interfacing—a layer of fabric placed between the garment fabric and its facing. Interfacings have always been an important part of garment production, especially for structuring and support in tailored jackets and coats. They are also used to reinforce details such as collars, lapels, cuffs, pocket flaps, buttoning areas, and waistbands.

Formerly, interfacings consisted primarily of linen, burlap, or horsehair. There have since been many developments, involving natural fibers, man-made fibers and blends.

Types of interfacings

Stable interfacings have no give and are designed to add stability to heavier tailored garments. Stable interfacings can be either woven or nonwoven textiles. As with other textiles, woven interfacings are made by interweaving yarns at right angles. *Nonwoven* interfacings have no grain and therefore can be cut in any direction. *Stretch-knit* interfacings have been developed for knit fabrics, but because interfacings are inherently intended to stabilize, they really do not give, especially when fused (bonded by heat to fabric). Stable and stretch interfacings are available in both fusible and sew-in forms. *Fusible* interfacings have become important labor and time savers.

Weights of interfacings

Basically, interfacings are available in three weight categories: light, middle, and heavy. The choice depends on the weight of the garment fabric as well as the support and effect desired. The interfacing should never be heavier than the actual garment fabric. Heavyweight fleece and thermal interlinings are also used in jackets and coats to give additional support and warmth.

Narrow Fabrics

Woven, knit, and braid trims are produced by narrow fabric manufacturers. Narrow fabrics include narrow laces, ribbons, braids, other woven and knit decorative bands, pipings, and cordings. Formerly, most of the trims sold in the United States were imported from Europe. Now, only the finest trims to use on high priced fashion are still imported from Europe. Some imports come from Asia but the domestic narrow fabrics industry has been able to compete.

Narrow fabrics are either functional (as in the case of tapes for inner construction) or decorative. The decorative trims industry is a fashion-oriented business, although there are really no seasonal lines. The fashion for trims is cyclical, like any other fashion. When an ethnic look is popular, trims are in great demand. Trims are used most frequently for children's wear and less for women's and men's clothing.

Trim types

There is a great variety of trim types, specially made for different uses and methods of application:

Appliqué. An individual motif that can be attached singly or in multiples as a trim. This term is also used for a technique to attach the edges of a motif to fabric.

Banding. Narrow fabric having two straight or decorative edges that make it interchangeable for various trims such as borders, edging, insertion, or accenting of a design line.

Beading. Openwork trimming, usually lace or embroidery, through which a ribbon may be threaded. (This beading is not to be confused with beads used to create banding or edging for evening wear.)

Binding. Prefolded trim that encloses a raw edge, finishing and decorating it at the same time.

Edging. A trim having one decorative edge and one straight edge.

Galloon. Lace, embroidery, or braid having two shaped-and-finished edges. It can be used as a banding, border, or design-line accent or can be applied just like insertion for a see-through effect.

Insertion. A trim with two straight edges (also a method of applying trim by inserting and sewing it between two cut edges of fabric).

Medallions. A chain of trims with individual motifs that can be used continuously as banding, edging, or galloons, or clipped apart and used as appliqués.

Laces

The popularity of handmade laces led to the invention of lace-making machines. Although narrow-textile manufacturers primarily produce bands of lace, their methods can be applied as well to the production of lace fabric. Laces are used primarily for lingerie or bridal gowns, but like other decorative trims, they recurrently play an important role in fashion. Today, there are basically four types of machine-made lace.

Barmen lace has its roots in Germany and Cluny, France. A Jacquard system governs the movements of the Barmen machine. Threads on bobbins are plaited together to resemble heavy, crocheted lace.

Leaver lace, named after Englishman John Leaver who developed the machine, is sheer bobbin lace. Also Jacquard-programmed, the Leaver machine twists threads into a giant web that can have up to 180 bands, each connected by auxiliary threads. To separate the bands, one of two methods is used. Auxiliary threads can be hand-pulled (as is usually done in Europe), or certain threads can be dissolved in an acetone solution.

Raschel knitted lace is less expensive than Leaver because of the high speeds at which it can be produced. This lace is also made in a giant web formed by linking chains of yarn, however, the bands are separated by pulling drawstrings.

Venice lace is made on Schiffli (see the following section) embroidery machines. The needles are controlled by a Jacquard system.

Embroidery

Although embroidery is not confined to narrow fabrics, it seems appropriate to discuss its production after that of laces. As an all over or as a trim, embroidery has been used most often in lingerie. Recently, though, it has become popular to use small single embroideries on the body of a garment, especially since machines have been developed to mass-produce motifs.

A computer automated embroidery machine.
(Courtesy of Sauer Textile Systems, Charlotte and Switzerland)

Schiffli is a continuous embroidery used on fabrics or for trims. Often a manufacturer will send out fabric to have it embroidered and then insert it into a garment as a panel, as on a blouse front. Schiffli embroidery machines are huge and highly automated but require some labor to operate the machines, sew sections together, do hand detailing or cut and fasten threads. Schiffli machines are also used to make embroidered eyelet (cut-work embroidery) trim or fabric. Some embroidery work is moving south to the Carolinas, other embroidery is done offshore.

Framework embroidery is used when a single motif is needed on a garment. Multi-headed machines, operated by a computer, embroider many pieces at one time.

Ribbons

Ribbons are another category of narrow fabrics. Many ribbons are woven at one time on looms that have 2, 4, 6, 8, 16, 24, or 48 spaces. The width of the ribbon and the volume of production determine the size of machine to be used. Satin and velvet ribbons are made much like their fabric counterparts. Grosgrain is woven on a belting or dobby loom.

Ribbons can be yarn-dyed or solid colors can be dyed after weaving to match manufacturers' color specifications. However, patterns such as stripes are woven with pre-dyed yarns. For intricate patterns, Jacquard looms are used. Some motifs, such as polka dots, are printed on the woven ribbons.

Although most ribbons are produced domestically, fine ribbons are imported from Europe: velvet from France, Germany, and Switzerland and plaids from England. Offray is the largest manufacturer of ribbons in the United States.

Throwaway or *craft* ribbon is woven on a broadloom and cut into bands. This type of ribbon is usually reserved for the gift-wrapping market. However, if the edges are embossed, it may be used in less expensive garments and accessories.

Braids

This category includes not only braided trims but also woven and knitted flat, decorative trims that are heavier than ribbons.

Braided trims are made by interlacing three or more yarns to form a flat narrow fabric. They include soutache, middy braid, braided cord edging, and rickrack. Knitted braids, made by a few needles on a warp-knitting machine, include both flat and fold-over braids. Woven braids, made on textile looms, include flat and fold-over braids and piping made with bias strips. Patterned braids, like ribbons, are produced on Jacquard looms.

Hook-and-Loop Fasteners

One of the most widely used fasteners today is hook-and-loop tape. It is used to fasten apparel and shoes, to secure shoulder pads, and for many other applications. In the 1940s, the Swiss inventor George de Mestral came home from hiking with cockleburs on his trousers. His microscope showed that cockleburs were covered with hundreds of tiny hooks and that his trousers were made of hundreds of tiny loops. De Mestral decided to turn his discovery into a practical fastener. He offered his new product under the trademark Velcro from the French words *velour* (velvet) and *crochet*. At first Velcro was very expensive and limited in its use. The first application in apparel was in skiwear.

In the apparel industry, most hook-and-loop (or touch) fasteners are made of woven or knitted nylon. The hook-and-loop tapes come in many stock colors or can be custom dyed. They are sold in 50-yard or 100-yard rolls in a variety of widths; 5/8 inch and 3/4 inch are the most common for apparel. A wide variety of fasteners is now available including linear tapes and cut pieces, straps, and custom forms. For example, to eliminate sharp edges for children's wear, the fasteners are often dye-cut with rounded corners and applied by sewing.

Zippers

Whitcomb Judson, a Chicago inventor, first introduced a metal "slide fastener" in 1891. However, not until several developmental stages later could a practical fastener be made on a tape. In 1923 the B. F. Goodrich Company first used the term *zipper* to describe the fastener, and manufacturers began to use the device in fashionable garments. In 1960 the nylon zipper was introduced, which provided the industry with an alternative lightweight zipper that could be dyed to match any garment fabric. Zippers are not only functional but can be decorative as well. Fashion sometimes calls for zippers to be shown on purpose.

Zipper producers are losing business because of the huge increase in apparel imports with zippers supplied by Asian manufacturers. There are several methods of zipper production, depending on the materials used. Zippers made from interlocking metal scoops (teeth) are either clamped or hot-molded onto presewn tapes. For nylon zippers, a continuous filament of nylon is first molded into a coil. Then the tape is either woven around the coil or the coil is sewn onto the tape. The chain of scoops or coil ranges from a very fine size 1 to a bulky size 9. After zipper lengths are cut to various size specifications, a lock is stapled at the bottom and stops are stapled to the top. Finally, a slider is attached that opens and closes the zipper. (At present the 7-inch, 9-inch, and 22-inch fasteners outsell most other sizes for use in skirts, pants, and dresses.)

Environmentally friendly, non-electroplated hardware made from non-rusting metal alloys such as stainless steel, nickel silver and brass used for Esprit's Ecollection. *(Courtesy of Esprit)*

Buttons

Buttons are an important aspect of a garment both for functional purposes and for fashion interest. When fashion emphasizes detailing, the market becomes stronger for buttons. Most button lines follow fashion trends in both color and styling. Currently, the button business is booming. Designers are using the buttons as a fashion statement, mixing sizes, colors, and shapes in one garment.

Buttons are both imported and domestically produced. New York City was originally the center of button production in the United States. Now, although the markets remain there, much of the production has moved in search of lower costs for labor and facilities.

Traditionally, buttons were made from natural materials such as wood, metal, pearl, porcelain, and bone. Manufacturers concerned with the environment are again promoting the use of natural materials. However, due to technological advances and the wide use of synthetics, less than 10 percent of all buttons sold in the United States are constructed of natural materials. Most buttons are now made of polyester or nylon because of the versatility of these materials and because they can be dyed to match fabric. Polyester is produced in sheet form and the button shapes are cut out of the sheet. In the case of nylon, liquid nylon is poured into button molds.

Most metal buttons are made from brass; a stamped-out face is put together with a stamped-out back. There are also some pewter buttons, which are cast. Metalized buttons are made with a core of acetate, which is covered by a coat of copper and finally a gold- or silver-look metal finish. Some environmentally concerned manufacturers are using buttons made from nonrusting alloys to avoid the

Button size chart.
(Courtesy of Blumenthal & Co., Inc.)

hazardous sludge which is a byproduct of electroplating. Most of these buttons are produced in Connecticut. (Other metal garment and accessory fasteners include buckles, eyelets, and rings.)

Mother-of-pearl buttons are made from natural shells found in the waters around Japan, the Philippines, and Australia. Relatively expensive, they are often copied in plastic. Other materials used for buttons, such as leather, wood, and bone, are expensive and are also copied in plastic.

There are basically two types of buttons: sew-through (cheapest to apply to garments) and shank (with a loop on the back). Buttons are purchased by the gross (144), and all sizes are expressed in terms of *lignes* (one ligne equals 0.025 inch or 0.635 millimeter). For example, a 1-inch button is the same as a 40-ligne button. Self-covered buttons are made to order for apparel manufacturers by contractors.

Belts

There are two types of belt manufacturers. The *rack* trade (see Chapter 11) sells to retailers, and the *cut-up* trade sells less expensive belts to apparel manufacturers to be used on dresses, jackets, skirts, or pants.

Garment belts are usually made of inexpensive materials such as bonded leather, vinyl, or ribbon. The belt material is either glued or sewn onto a stiff interfacing or backing. Stitched belts hold up longer than glued ones. Novelty belts can be made of anything needed to meet fashion requirements: webbing, plastic, braid, chain, rope, or even rubber.

Belts can also be made in house by the garment manufacturer with buckles from the button supplier. Most of these belts are made from garment fabrics. Self-belts can also be made to order by a contractor.

Labels

Labels are the sources of identification for garments. They tell the consumer the name of the manufacturer and sometimes the designer. Some people are attracted to a particular manufacturer because of its advertising or the quality or fit of its garments. A label is thus very important to them.

The label itself can help establish the company's image. A large woven label on satin ribbon denotes a higher-priced designer garment, whereas a small printed label usually signifies a less expensive one.

Woven labels are produced on a Jacquard loom—either a conventional shuttle loom (that makes ribbon with finished edges) or on a broad loom (the fabric is then cut into strips with hot knives that fuse the edges). The broadloom has the advantage of offering a wider assortment of colors; whereas only two can be used on a shuttle loom. Labels are usually made on satin or twill and are sewn at the centerback of the neck, on the waistband, or on the inside jacket facing. There are four types of folds for hiding the ends of the ribbon. The choice depends on the shape of the label, its placement, and the sewing method. Pressure-sensitive labels have also been developed, but are not used extensively in the fashion industry.

A label showing fiber content, care instructions, and country of origin must also be included in the garment. This information can be part of the primary label but is usually separate, just underneath the manufacturer's or store label.

There are pages of government regulations concerning fiber-content and country-of-origin labeling. The predominant fiber in the garment is listed first along with its content percentage, followed by the other fibers in decreasing order of content. Care instructions are recommended and can be very specific. These instructions not only protect the consumer but also protect the manufacturer against returns made by consumers who do not care for their garments as instructed! (Country-of-origin labeling is discussed further in Chapter 10.)

Care labels can be printed, or made in house on a Soabar machine. The Soabar machine allows changes to be made quickly as needed for smaller quantities.

Hang tags, if used (they are not required), are the last item of identification added to a garment. They are used primarily as size identification or just to draw attention to the garment, fiber, or fabric producer. Hang tags are supplied by the fiber or fabric producer and hung on threads from the side of the garment, so that the customer sees them first when looking at garments sideways on a hanger rack.

THE LEATHER INDUSTRY

Designers, merchandisers, and manufacturers of leather apparel and accessories must know the properties and best uses of leather. Retailers, too, are interested in the quality of the product they sell.

The many varieties of leather produced throughout the world are used in hundreds of different ways. Leathers are preserved animal hides and skins, byproducts of the meat industry. The world's largest producers of cattle hides are Russia, the United States, and India; most goatskins come from India and China; and most sheepskins from Australia, New Zealand, and the republics of the former Soviet Union. The United States imports almost as much leather as it exports. Developing countries, particularly those with abundant raw material supplies such as Argentina, Brazil, and India, impose export controls or taxes in order to encourage the growth of their own tanning and leather products industries.

Due to competition from cheap labor in developing countries, the tanning industry in the United States has shrunk to only 110 facilities clustered in Wisconsin, New England, and Tennessee.[1] *Tanneries* purchase and process skins and hides and sell the leather as their finished product. Companies usually specialize because processing methods depend on the nature of the skins treated and the end use of the leather.

Hides and skins are differentiated by weight. *Skins* come from smaller animals such as goats, deer, pigs, and calves, and weigh less than twenty-five pounds. *Hides* come from large animals such as steers, cows, buffaloes, and horses, and weigh over twenty-five pounds each.

Leather Processing

The process by which hides and skins are made into leather can take up to six months and requires extensive equipment and skilled labor. There are three basic steps in leather processing: pretanning, tanning, and finishing.

Men's fashion in leather by Giorgio Armani. (*Courtesy of Giorgio Armani, Milan, Italy*)

Pretanning

Pretanning is basically a cleaning process. The cured hides are soaked in water to rehydrate them and to remove dirt, salt, and some proteins. Hides and skins must also have hair and fat removed by additional treatments, which differ with each type of hide or skin.

Tanning

Tanning involves the application of various agents that protect the hide or skin against decomposition. Treatment methods include soaking and powdering. The choice of agents depends largely on the end use of the leather. Tanning agents include vegetable products, oils, minerals, and chemicals. The roots, bark, wood, leaves, or fruits of various trees and shrubs provide the vegetable tannins, which produce a firm, heavy leather. Oil tannages, especially codfish oil, produce relatively soft and pliable leathers such as chamois. Mineral tannages, such as alum and especially chrome, are common, as are the newer chemical tans, such as sulfonic acid. Each tanning agent imparts special characteristics to leather. A combination of agents usually produces the best results.

Finishing

This step produces the desired thickness, moisture, and aesthetic appeal. After excess water and wrinkles are removed from the tanned leather, it is shaved or split into uniform thickness and then dyed. Color is applied to leather by brushing, or by tray, drum, spray, solvent, vacuum, or tank dyeing. Special effects may be created by dabbing on color with a sponge, padding color on through stencils, sprinkling, spraying, or tie-dyeing. The dyed leather is treated with oils and fats, which provide lubrication, softness, strength, and waterproofing. The leather is then dried to fix the dyes and oils permanently. Dried leather is conditioned with damp sawdust to obtain uniform moisture content and then stretched for softness. Finally, the leather is coated with a seasoning or finish that improves its properties or character. For example, urethanes are used to add shine to patent leather. Leathers may be further treated with buffers, rollers, or presses to achieve glazed, matte, and embossed effects.

Examining a leather skin.
(Courtesy of Bally of Switzerland)

Environmental Concerns

The Environmental Protection Agency (EPA) has established standards to control the polluting wastes that tanners discharge such as sulfides, chromium, and acid. Control of these wastes requires expensive primary and secondary treatment facilities. The industry is developing and adopting new tanning systems that will use nontoxic metal salts and other organic tanning materials to replace the chromium. The industry is also encouraged to adopt low-solvent or solvent-free finishes.

Leather Promotion and Marketing

Like the fiber and fabric producers, the leather industry promotes its products to apparel manufacturers, fashion editors, retailers, and consumers. Promotion is done by the company or through trade associations such as the Tanners Council. Yet individual tanners, unlike fiber producers, are not known by the public. Retail advertising for leather products may mention the type of leather and the designer, but not the producer.

Leather markets are concentrated in Western countries. The major exporters of finished leathers are Japan, South Korea, Taiwan, Italy, and Canada.[2] International trade in leather is over $1 billion per year.

Rising world population and incomes, along with new fashions for leather, have contributed to an ever increasing demand. However, the use of synthetics as leather substitutes is growing. In Western countries, man-made products have supplanted 50 percent of the goods formerly made of leather. To protect and expand their markets, leather producers must constantly strive to develop new leather types and colors.

THE FUR INDUSTRY

The fur industry is another important supplier of raw material for the fashion industry. Many apparel designers also design collections for fur manufacturers.

Fur is the hairy coat of a mammal. However, except in the case of lamb, fur is not a by-product of the meat industry. From prehistoric times, people have used animal fur for both its warmth and its attractive appearance. Because fur has long been associated with wealth and prestige, the demand for luxury has played a major role in the development of the fur industry.

This section discusses the characteristics of various furs, the processes involved in preparing fur pelts for their use in the manufacture of fur garments, and the marketing of furs.

The fur industry consists of three major groups: pelt producers or trappers, fur processors, and companies that produce fur garments for consumers (fur garment production is discussed in Chapter 11).

Fur Sources

The industry must first obtain the pelts or skins of fur-bearing animals. Furs come from wild animals and, more often, fur farms or ranches.

The international fur market is dominated by mink production. Scandinavia produces 45% of the world mink supply, Russia 31% and the United States 10%.[3] The major animals raised or trapped for the United States fur business are (in descending order of importance): mink, mostly bred; fox, trapped and bred; and sable, imported from Russia and Canada, where it is trapped. Other furs, used in smaller quantities, include muskrat, skunk, opossum, lamb, and rabbit.

Wild furs

Wild furs for commercial use come from over eighty countries on all six continents, but mainly from North America, which has the greatest variety (forty different types). In the late 1960s individuals, international organizations, and governments became concerned about the possible extinction of endangered species. As a result, some countries enacted legislation restricting or prohibiting the commercial use of particular animals, including certain monkeys, seals, and leopards. Wild fur-bearing animals are usually caught in baited traps. There is public concern about the agony animals suffer in this form of slow, painful death. There is new legislation against certain kinds of traps to try to make trapping more humane. In some areas seals are clubbed to death, a practice that has caused protests and efforts to prohibit the commercial use of wild seal fur.

Fur farming

Fur farming has greatly increased the supply of fur for the fashion industry. General livestock methods are adapted to the keeping and breeding of animals for their pelts. Research in management, feeding, and breeding techniques has resulted in the production of quality furs in thousands of mutations. Mink has become the most popular fur, accounting for approximately 60 to 70 percent of today's fur trade. Silver fox is also raised mainly on farms.

Inspecting fur pelts.
(Courtesy of the Deutsches Pelz Institut, Germany)

Marketing

Trappers normally sell their catches to collecting agents, who in turn sell them at auctions or to wholesale merchants. These merchants maintain stocks of furs, selling them to manufacturers as they are needed.

Fur farmers and ranchers, however, sell their pelts at public auction directly to the wholesale pelt merchants or manufacturers, or to commission brokers (who buy for merchants or manufacturers). Major fur auction centers are located in New York City, Frankfurt, St. Petersburg, and Montreal. Pelt prices can fluctuate sharply, because price is dictated by supply and demand.

Fur processing

After manufacturers purchase pelts at auctions or from wholesale pelt merchants, they usually contract with fur dressing and dyeing firms to process them. New York City is the largest fur-processing center in the United States.

Dressing

Fur skins are dressed to make them soft, pliable and lighter weight as well as to preserve their natural luster. Dressing processes vary with the nature and condition of the skin, but there are usually at least four distinct steps: preliminary cleaning and softening of the pelt; *fleshing* (cleaning) and stretching; *leathering* (a tanning process using oils or other solutions); and finishing. The industry would like to make the process less toxic.

Dyeing

Dressed pelts may be sent out to a dyer. The modern use of chemical compounds known as *fur bases* has enabled fur dyers to produce a wider variety of colors. Although not all furs are dyed, dyeing has led to the use of many skins that were unattractive in their natural colors. New colors are an important selling point among fur retailers. Dyers often keep their techniques secret from competitors to maintain a market edge. Following the dyeing process, the pelts are now ready for garment production, a process covered in Chapter 11.

SUMMARY

Trimmings are the materials needed to finish and decorate garments and accessories. The decorative materials include buttons, ribbons, laces, braids, and belts. The functional materials, often classified as findings, include thread, interfacing, zippers, tapes, and elastic. Thread, interfacing, and narrow fabrics are an extension of the textile industry.

Thread is what holds garments together. Woven and nonwoven interfacings are needed to give support and shape to garments. Narrow fabrics, which can be woven, knitted, or braided, include lace, ribbon, braid, piping, cording, tapes, and banding. The primary types of machine-made lace today are Barmen, Leaver, Raschel, and Venice.

Zippers, buttons, and belt buckles are the hardware used in garment and accessory manufacture. There are two basic kinds of zipper construction: either metal scoops are clamped onto tapes or tapes are woven around a nylon coil. Buttons, formerly only made out of natural materials such as wood, metal, pearl, porcelain, and bone, are now primarily made from nylon or polyester for commercial use. Fabric belts for the cut-up trade can be sewn or glued onto stiff backings and finished with buckles. Each type of trimming is a separate industry with its own resources and markets and is a study in itself.

Leather and fur, the oldest body coverings, are still an important part of the fashion industry. Animals, either wild or raised on farms, are the sources of both leather and fur. Tanning (for leather) and dressing (for fur) are similar treatments: they clean, preserve, and bring out the natural beauty of the skin, hide, or pelt. Although modern technology has speeded processing and made it somewhat easier, leather and fur production are still basically craft industries.

CHAPTER REVIEW

Terms and Concepts

Briefly identify and discuss the following terms and concepts:

1. Thread
2. Uses of elastic
3. Stable interfacings
4. Narrow fabrics
5. Barmen, Leaver, Raschel, and Venice lace
6. Examples of braids
7. Zipper scoops
8. Coil zippers
9. Button materials
10. Cut-up belt trade
11. Leather types
12. Tanneries
13. Steps in processing leather
14. Fur dressing

Questions for Review

1. Briefly explain the difference between decorative trims and findings.
2. Discuss the various types of interfacings and their uses.
3. What categories are included in narrow fabrics?
4. Describe the four basic kinds of machine lace.
5. Describe the two basic types of zippers and their construction.
6. Discuss the various natural and man-made materials used to make buttons.
7. Explain how belts are produced for the cut-up trade.
8. Briefly discuss the three basic steps in leather processing.
9. Discuss the animal rights controversy and how it might affect the fur industry.
10. Discuss the four steps in the dressing of furs.

Projects for Additional Learning

1. Select one category of trim (lace, for example) and find a variety of samples at fabric stores. Ask if you may have or purchase just a small piece of each. Describe the pattern or construction of each type. Show examples (photos or actual garments) of their end use.
2. At the library, trace the origin and development of a specific trim. Explain the differences between how it was made by hand originally and how it is mass-produced today.
3. Visit a local department or specialty store and find five garments that use decorative and functional trimmings. List every trim on each garment. Discuss how the trims affect the total design of the garment. How do you think the manufacturers balanced the added cost of the trimmings with fabric and labor costs?
4. Write to a national animal-protection agency such as API (Animal Protection Institute) or PETA (People for the Ethical Treatment of Animals). Find out what members are doing to promote animal rights. What impact do prevailing political attitudes and economic conditions have on the sale of furs?
5. Visit a furrier or fur dealer. Learn to identify various furs by their characteristics.

NOTES

[1] Charles Myers, Leather Industries of America, Interview, February 5, 1993.

[2] "U.S. Industrial Outlook 1993," U.S. Department of Commerce, January, 1993, pp. 33–34.

[3] "Mink Market Share," Fur World, November 26, 1992, p. 5.

Part Three

THE MANUFACTURING OF FASHION

Christian Dior salon on the corner of rue François Premier and avenue Montaigne in Paris.
(Courtesy of the House of Dior)

8

INTERNATIONAL FASHION CENTERS

CAREER FOCUS

Every professional in the fashion business wants to be informed about the newest trends in fashion. To do this, they constantly read newspaper articles that tell what is happening in the fashion capitals around the world. Manufacturers and buyers want to know what the trendsetting designers are showing in their latest collections.

CHAPTER OBJECTIVES

After reading this chapter you should have attained competence in the following areas:
1. Knowledge of the names of well-known international fashion creators.
2. The ability to explain the reasons for French fashion leadership.
3. The ability to describe the growth in importance of the prêt-à-porter.
4. Understanding the reasons for the importance of New York as a fashion center.
5. The ability to outline the role of international and domestic fashion centers.

This chapter introduces the major fashion capitals of the world, the centers that are most influential in creating, manufacturing, and marketing new fashion. You will read about the specialties of each and about the creators who have made them into fashion centers. Fashion centers develop because of concentrations of resources, supplies, skilled labor, and creativity. All designers are influenced by what other designers and artists are creating. Excitement about a new idea acts as a catalyst for more creativity. That is why many creative people gravitate to major creative centers.

Three cities have emerged as major fashion capitals: Paris, France; Milan, Italy; and New York City in the United States. Other noteworthy but less influential centers include London, England; Tokyo, Japan; Madrid and Barcelona in Spain; Düsseldorf and Munich in Germany; Montreal and Toronto in Canada; and Los Angeles and San Francisco in California. American designers score high when it comes to marketing savvy and making saleable clothes that appeal to the whole U.S. population. However, retailers still believe that "real creative originality is abroad."[1]

However, it is increasingly difficult to describe the characteristics of fashion by country or fashion capital. As President Clinton said in his inaugural address, "There is no longer clear division between what is foreign and what is domestic." The fashion industry is becoming a worldwide exchange of ideas, talent, material, and products. Ideas come from all over the world, textiles are exported from one country to another; production is done almost everywhere, and nearly every country contributes in some way.

Designers work or show their collections in other countries. Karl Lagerfeld, a German, works in Paris. Should we consider him a German designer or a French designer? Gianfranco Ferré, an Italian, designs the Dior couture collection in Paris. Keith Varty and Alan Cleaver, Englishmen, design for Byblos in Italy. Rifat Ozbek, from Turkey, works in London but shows his collection in Milan. The list is endless.

The job of the international designer is not an easy one. Ferré commutes between Milan and Paris. Oscar de la Renta commutes between New York, Paris, and his home in Santo Domingo. Lagerfeld designs five collections: his own signature line, Chanel couture, Chloe, a less expensive KL line produced in Germany, and furs for Fendi of Italy. Some designers have time for little else than designing or promoting those designs. They supervise large design teams for several collections, visit factories, attend store openings, and make worldwide public appearances.

Their names become famous because of extensive press coverage, their own promotional efforts, and success in the marketplace. With fame come requests for *licensing,* a process through which manufacturers are given permission to use the designer's name. Their fashion empires grow to the point that successful international designers have become stars.

A successful fashion business is not built on a good designer alone. The top designers have astute business partners like Pierre Bergé at St. Laurent, Giancarlo Giametti at Valentino, or Peter Strom at Ralph Lauren.

One trend in the fashion business is to try to build powerful conglomerates. Bernard Arnault owns Dior, Givenchy, and Louis Vuitton, and has also backed Lacroix. Company ownership has become so complicated that it is often difficult to judge the origin of a collection. The Cerruti 1881 line, for example, originates in Italy, but the production and distribution rights have been purchased by Escada, a German firm. The production and distribution rights for the American Ralph Lauren Womenswear collection is owned by Bidermann Industries of Paris.

FRANCE

Paris has long been the foremost city of world fashion and it has managed to keep its position in spite of fierce competition.

Paris is the capital of France and the Hollywood of the fashion world. Fashion is one of France's top three export industries and the second most important in employment with 370,000 workers. Paris became the capital of fashion because it had the necessary resources and a creative atmosphere. In addition, the French government has always supported and encouraged *les mains de France* (the hands of France), giving the needle trades much deserved respect. There is also tremendous cooperation among the French design firms, fabric mills, and the auxiliary shoe, hat, fur, trimmings, findings, and embroidery industries. A designer who needs special fabrics finds the mills willing to weave or print just a few meters as a test run. Shoe manufacturers plan designs to complement designer garments, and button and trim manufacturers will create items for the exclusive use of one designer. Oscar de la Renta describes the situation as "the extraordinary support system and all the little artisans that make Paris special."[2] Having such an atmosphere, Paris is understandably looked to for fashion leadership.

Recognizing Paris as a fashion center, many fiber and fabric associations, promotion agencies, and information sources have established their main fashion offices there. Among the designers from other countries who are now showing their collections in Paris are

Valentino from Rome, Kansai Yamamoto from Japan, and Katharine Hamnett from London. Paris attracts talent from around the world, which in turn keeps Paris the center of fashion.

The Couture

"Fashion is a very important economic sector for our country, and couture is the flagship of French fashion and the radiance of France," explains Dominique Strauss-Kahn, the French Minister of Industry.[3] *Couture* is simply the French word for fine custom dress design, made to measure for a particular customer. A *couturier* is a male couture designer, a *couturière* his female counterpart. A couture business is called a *maison* (house). The private client comes to the *salon*, an elegant showroom in the same building as the design studio, to see sample garments in the collection. When a client orders a dress or suit, it is made up in her exact measurements, with several fittings. Construction usually takes weeks. *Haute couture* (the most exclusive couture) is reserved for the very best design and highest quality of fabrics and workmanship. However, many people today find that couture is a dying art. Fewer than a thousand women in the world can afford to spend $10,000 to $20,000 for a suit or up to $50,000 for an elaborate gown!

It costs approximately $1 to $2 million a year for a large house to produce its couture collections. Collection costs include fabrics, labor, especially made accessories, expensive models, catwalk, sound systems, rent of a theater, and dinners for buyers. However, the French government gives support, in the way of low taxes and tax credits, to the couture in order to promote exports. Government-owned French television gives the

Italian designer Gianfranco Ferré fitting a toile at Christian Dior in Paris where he designs the couture collection. *(Courtesy of Gianfranco Ferre)*

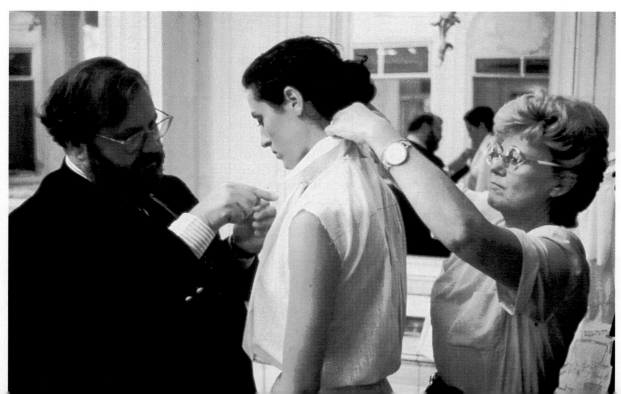

couture free exposure. This kind of publicity is more prestigious than costly advertising and helps to compensate for the costs of producing a collection. The publicity is especially important because it generates sales of ready-to-wear, perfume, and licensing businesses, thereby helping to keep the couture alive.

The couture is regarded as offering the opportunity for the purest form of creativity in fashion, providing the "research and development" for the French fashion industry. Emanuel Ungaro explained, "We have the freedom for creation in couture...It is our duty to serve as a laboratory for fashion and a guide to ready-to-wear."[4] The clothes of some couturiers are audacious but this is usually done to attract attention. Other couturiers feel that it is their responsibility to make pretty clothes that are flattering and saleable. Christian Lacroix remarked, "Haute couture runs the risk of being suffocated by its own traditions. If it is to survive, it must be firmly anchored in daily life."[5]

Parisian Couturiers

Karl Lagerfeld brought international attention back to the couture when he took over as artistic director for the house of Chanel (shah-nell´) in 1983. He also designs the Karl Lagerfeld, Chloe, and KL ready-to-wear collections, as well as furs for Fendi of Italy. Lagerfeld won the International Wool Secretariat design competition in 1954 at age 16 and was hired by Pierre Balmain as an assistant. He was designer for Chloe for 19 years before Bidermann Industries offered to back him in his own line.

German designer Karl Lagerfeld who works in Paris where he designs his signature collection, Chanel couture, Chloe, furs for Fendi of Italy, and the KL line for Steilman of Germany. *(Self portrait courtesy of Karl Lagerfeld)*

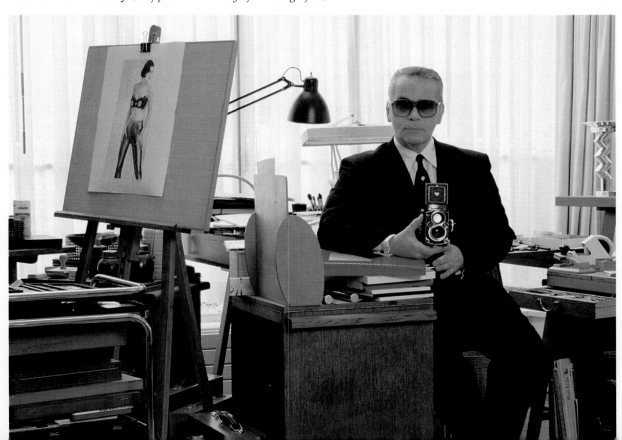

Because he is the major international trendsetter, he is often referred to as "King Karl" by the fashion press. "...When Karl speaks, the fashion world listens."[6]

Yves Saint Laurent (eve sahn´ law-rahn´) is still considered a master of couture. He is a steady, major influence in the fashion world, setting trends in a restrained, sophisticated way with a good sense of timing. He began his career by winning the wool design competition at age 17 and a post as assistant to Dior. He opened his own business with Pierre Bergé in 1962 at the age of 26 and now has over 200 licenses. The Metropolitan Museum presented a retrospective of his work in 1984.

Christian Lacroix (la-kwa´), who also helped to bring excitement back to the couture, is known for his extravagant costume looks, silhouettes, colors, and pattern mixes. Born in Arles, France, in 1951, he studied art history, intending to become a museum curator. Chance brought him to the fashion house of Hermès in 1978, and he became the artistic director of Jean Patou in 1981. In 1987 Bernard Arnault, president of Financière Agache, offered to back him in his own couture house.

Gianfranco Ferré for Dior (fer-ray´), an Italian, was chosen by Bernard Arnault to replace Marc Bohan at the house of Dior in 1989. He does his own ready-to-wear line in Milan (see Milan).

Emanuel Ungaro was born in 1933 and learned his trade in his father's tailor shop. At the age of 22 Ungaro left for Paris; he eventually spent four years with Balenciaga and a year with Courrèges. He opened his own salon in 1965 and added a men's wear collection in the late 1970s.

Hubert de Givenchy (ghee-vahn-shee´), born in 1927, studied at the École des Beaux Arts (school of fine arts). At age 17, he started designing at Fath, then at Piguet, Lelong, and Schiaparelli. He opened his own house in 1952 at the age of 25, when he was especially known for making clothes for Audrey Hepburn. He is respected for the refined elegance of his creations.

Oscar de la Renta for Balmain, a well-known American designer, has joined the ranks of the international commuter designers to design couture in Paris.

Other couturiers include Michel Goma for Balenciaga, Pierre Cardin, Carven, Tom Van Lingen for the reopened house of Jacques Fath, Louis Feraud, Lanvin, Oliver Lapidus for Ted Lapidus, Michel Klein for Guy Laroche, Hanae Mori, Patou, Paco Rabanne, Gerard Pipart for Nina Ricci, Rochas, Erik Mortensen for the house of Scherrer, Per Spook, Torrenté, and Philippe Venet. Also Italian designers Valentino and Versace show their couture collections in Paris.

In France, only those on the selective Couture-Creation list of the Fédération Française de la Couture are considered members of the haute couture; membership is based on high standards of excellence and special requirements. The federation includes three branches: couture, women's ready-to-wear, and men's wear. Under their auspices, the Chambre Syndicale dictates rules about workrooms and collections.

Ateliers

The couture salons, *boutiques* (retail shops for ready-to-wear and accessories), and *ateliers* (sample workshops) center in the avenue Montaigne and the Faubourg Saint-Honoré in Paris. To qualify for the Couture-Creation list of the Chambre Syndicale de la Couture Parisienne, a house is required to have at least one atelier in Paris with a minimum technical staff of 15, not including the director. A couturier may have anywhere from 20 to 400 employees in one or a combination of several ateliers. In order to get established, novice couturiers are allowed a two-year transition period during which they need have only 10 employees.

In large houses, where there is much work to be done, *modelistes* work under the head designer, executing the individual designs. The modelistes are the liaison with the ateliers, where they supervise construction of the *toile,* or sample garment.

Ateliers are separated into *flou* for dressmaking and *tailleur* where they specialize in the tailoring of suits and coats. Each atelier is headed by a production manager or chief technician called the *premier d´atelier* (pre-me-ay´ dah-tel-yay´). Under the manager are the shop assistant(s), fitters, *midinettes* (seamstresses, classified as first hands or second hands), and apprentices.

In houses where the original couturier is no longer living, former assistants or other designers have taken over design responsibilities such as Ferré for Dior. Some houses have one designer for the couture collection and another for the prêt-à-porter line.

Prêt-à-Porter

Prêt-à-porter (pret-a-por-tay´) is French for "ready-to-wear." Most French fashion is mass-produced, as it is elsewhere. Mass production makes fashion less expensive. In this case, identical garments are manufactured in various sizes and colors. The customer needs only to try one on in a store to see if it fits, purchase it, and take it home immediately.

All the couturiers began to produce prêt-à-porter collections because the couture was not financially successful in the 1960s and 1970s. Today their prêt-à-porter creations cost as much as couture used to. Whereas a couture creation might now cost from $5000 to $50,000, the upper end prêt-à-porter range is roughly $500 to $2000. However, despite its price, a prêt-à-porter garment brings no exclusivity with it. There is the same experimentation involved in creating original sample garments for a ready-to-wear collection as there is for a couture garment, although construction is based on mass production methods. Only the mass production of those samples makes ready-to-wear a profitable business.

Prêt-à-porter designers

Most of the couturiers also have prêt-à-porter collections.

Karl Lagerfeld is a major ready-to-wear trendsetter both with his signature collection and his Chloe collection.

Yves St. Laurent, Christian Lacroix, and **Emmanuel Ungaro** also have important prêt-à-porter lines.

In addition, there are designers who have built their reputations on ready-to-wear. Some of the most well known are:

Jean-Paul Gaultier, known as the *enfant terrible* ("terrible child" due to his extreme creations) of the French fashion world, began his career with Pierre Cardin and Jean Patou. He began his women's collection in 1978, adding a men's collection in 1984 and Junior Gaultier in 1988. He also enjoys designing costumes for films and Los Angeles presented a retrospective of his work in 1992.

Claude Montana created his own company in 1979. He designs both men's and women's collections produced in Italy, a leather collection, knitwear, furs, shoes and accessories.

Thierry Mugler began his company in 1974. He also has licensed collections of knitwear, activewear, men's wear and accessories. He is also a professional photographer and has designed costumes for the theater.

Additional prêt-à-porter designers include Azzedine Alaia, Anne-Marie Beretta, Barbara Bui, Jean-Charles de Castelbajac, Corinne Cobson, Ann Demeulemeester, Marithe and Francois Girbaud, Kenzo, Helmut Lang, Herve Leger, Lolita Lempicka, Martin Margiella, Popy Moreni, Bernard Perris, Myrene de Premonville, Georges Rech, Sonia Rykiel, Elizabeth de Senneville, Martine Sitbon, Sophie Sitbon, Chantal Thomass, and Dries Van Noten.

To reach a wider audience, most designers also have a less expensive *diffusion* line such as Ungaro's Emanuel line. There are also many less expensive prêt-à-porter lines shown in group exhibitions at the Porte de Versailles. Some of these companies such as Naf Naf, that do not publicize the names of their designers, have become world famous brands and influential for junior fashion.

ITALY

Italian fashion is very popular with Americans, for it suits a casual lifestyle better than the more extreme French fashion.

The Italian fashion industry and its influence on the world have grown enormously. Around 1940 only 30 fashion manufacturers operated in Italy, their production limited basically to men's wear. The Italians built an international fashion reputation on beautiful fabrics, styling, knitwear, leather goods, tailoring, and quality production. Fashion is now Italy's second biggest industry (next to tourism). Clothing and textiles have become Italy's biggest export after shoes. The Italian fashion industry is primarily devoted to ready-to-wear and accessories.

Milano (Milan) has become the center for *moda pronta* (ready-to-wear) because it is close to the fabric sources of Como, Biella, and Turino. Italy also has minor fashion centers in Florence and Rome as well as fashion and accessory companies scattered around the country.

Designers, manufacturers, and fabric companies work cooperatively, often as integral parts of a large vertical company. Many of these fashion companies are part of family-owned textile firms. They are able to invest heavily in the newest technology and spacious modern factories that help Italy maintain its reputation for high quality production. The largest Italian based apparel companies are both vertical and family owned: Benetton, Marzotto, Gruppo Finanziario Tessiler (GFT), and Max Mara.

Italian designers

Giorgio Armani, born in 1934, wanted to become a doctor. However, he gave up his studies and became assistant men's wear buyer for Rinascente, an Italian department store. Later he was hired by Nino Cerruti where he learned apparel production. In 1974, he created his first collection under the Armani label with partner Sergio Galeotti and added women's wear in 1975. His $200 million empire includes Emporio Armani (stores and a less expensive collection), A/X Armani Exchange (basics and jeans) and licenses. His tailored style set international trends in the 1980s.

Domenico Dolce and **Stefano Gabbana** began working together in 1982. Dolce learned his craft working in his family's clothing factory in Sicily. Gabbana studied graphics in design school. They met in 1980 while working for a Milanese designer. In 1985, they presented their first women's collection under the Dolce & Gabbana label and have since added men's wear, knitwear, lingerie and swimwear.

Gianfranco Ferré, born in 1944, studied architecture, and today is considered the architect of Italian fashion. He gave up architecture to design accessories and later presented his first collection of ready-to-wear under the Baila label in 1974. His first women's collection under the Ferré name was shown in 1978 and his men's collection was introduced in 1982. In 1989 he was selected to design couture for Dior in Paris but also continues his ready-to-wear collections in Italy.

Romeo Gigli (ro may o   gee lee), known for his *minimalistic* style, first studied architecture in Italy and then learned tailoring in New York City. He presented his first collection in 1983 and has since added men's wear. He now has a diffusion line called G. Gigli and has chosen to show his women's wear collection in Paris.

Valentino Garavani (who uses only his first

Giorgio Armani at the Emporio Armani shop in Milan. *(Courtesy of Giorgio Armani)*

name professionally) was born in 1932 and at age 17 started to work for Guy Laroche and Jean Desses in Paris. He opened his own house in Rome in 1960 and shortly thereafter Giancarlo Giametti became his business partner. Because he is Italy's most successful couturier, Valentino has been considered the main link with Paris and now shows his collections there. He is a fastidious worker and is known for his elegant collections.

Gianni Versace, born in 1946, studied design in hopes of a fashion career. He worked as a tailor's buyer and later designed for Complice, Genny, and Callaghan. In 1978 he set up his own women's wear business with members of his family and added men's wear in 1979. He continues to design for Genny and has started a couture collection which he shows in Paris.

Other internationally known designers and manufacturers in Italy include Laura Biagiotti; Nino Cerruti, who put Italian men's wear on the map; the Fendi sisters—Anna, Franca, Alda, Paola, and Carla; Salvatore Ferragamo's sons and daughters—Ferruccio, Fiamma, Massimo,

Giovanna, Fulvia and Leonardo who have expanded from shoes to other accessories and ready-to-wear; Alberta Ferretti; Donatella Girombelli is chairman of Genny Holding SpA which owns Genny, Byblos and Complice; Mariuccia Mandelli, who owns Krizia with her husband Aldo Pinto; Rosita and Tai Missoni who design elegant knitwear; Franco Moschino; Luciano Soprani; Sportmax; Englishmen Keith Varty and Alan Cleaver who design for Byblos; and Zegna men's wear.

UNITED KINGDOM

London has an international reputation for men's tailoring and has occasionally enjoyed the spotlight for trendy fashion for young people

London, the capital of England, U.K., has long been the respected world center for classic men's business attire because of its famous Savile Row tailors. Prince Charles still has his suits made at Gieves & Hawkes at No. 1 Savile Row, where his father ordered suits as a young man. Other famous tailors include H. Huntsman & Sons, Bernard Weatherill, Anderson & Sheppard, Turnbull & Asser, and Henry Poole (the oldest Savile Row firm).

A *bespoke* (custom-tailored) suit, completed in three or four weeks, may cost between $1000 and $3000. The trendier English men's designers include Paul Smith, Charlie Allen, and Roger Dack.

British fashion is a curious combination of tradition, eccentricity, and international fashion talent. The British have a fine international reputation for woolen country clothing, trench coats, and cardigans from such companies as Jaeger, Burberry, Aquascutum, Austin Reed, Marks & Spencer, and McGeorge, as well as fabrics from Liberty.

The young designers of the 1960s made London a fashion capital, but designers have had a difficult time maintaining that reputation since the fading of "swinging London." The 1970s gave the junior fashion world the British *Punk* look and English designers were popular in the early 1980s, but London has not been able to recapture its former glory. In fact, it is difficult to identify a British look anymore. Some of London's best-known designers are showing their collections in Paris or Milan. The British Fashion Council was formed to try to mount a single women's wear exhibition for London Fashion Week to strengthen the United Kingdom's position in the market.

The public appearances of Princess Diana have helped to promote the work of London designers. Well-known London *off-the-peg*

Gieves & Hawkes at No. 1 Savile Row, London.
(Courtesy of Gieves & Hawkes)

(ready-to-wear) designers include Alistair Blair, Caroline Charles, Jasper Conran, Paul Costelloe, Ellis Flyte and Richard Ostell, Timney Fowler, John Galliano, Katharine Hamnett, Betty Jackson, Patricia Lester, Ben de Lisi, Jean Muir, Rifat Ozbek, Jenny Packham, Arabella Pollen, Helen Story, Catherine Walker, Vivienne Westwood, and Richard Nott and Graham Fraser at Workers for Freedom. Couture designers include Murray Arbeid, Bellville-Sassoon, Bruce Oldfield, and Zandra Rhodes.

Britain's apparel industry employs approximately 300,000 people. North of Oxford Street in London, in the area around Margaret Street, lies the West End "rag trade" district, which supports a conglomeration of fashion suppliers, studios, and showrooms. Actual manufacturing, originally confined to London's East End, has now spread all over England. However, as in other countries, many manufacturers have garments produced in Italy, Eastern Europe, or Asia.

GERMANY

More and more buyers are adding Germany to their European shopping trips. A "Made in Germany" label has enjoyed a strong quality connotation and now is becoming known for styling.

Unlike France, Germany's fashion industry is decentralized. Apparel companies are located all over the country, with large centers in Munich, Berlin, Krefeld, Düsseldorf, and Hamburg. The industry employs over 200,000, the largest group of workers in the German consumer goods industry, in over 2500 firms.

German designer Wolfgang Joop in his design studio.
(Courtesy of Wolfgang Joop, photo by Bernd Isemann)

Internationally known designers and fashion brand names include Iris von Arnim, Willi Bogner, Hugo Boss, Escada, Wolfgang Joop, Rena Lange, Rene Lezard, Mondi, Caren Pfleger, Uta Raasch, Jil Sander, Klaus Steilmann (the largest women's clothing manufacturer in Europe), Strenesse, and Windsor.

SPAIN

Spain is seen as Western Europe's fashion frontier and Spanish designers are enjoying the international attention.

The Spanish fashion industry consists of about 5000 firms and approximately 150,000 employees. The burgeoning industry has little international business or merchandising experience as it has been run from small shops and factories. Members of the Cámara de la Moda Espanola are scattered throughout the country. One of the problems in promoting Spanish fashion is that half of the designers show in Madrid and half in Barcelona. Efforts are being made to try to consolidate government-sponsored presentations of men's wear in Barcelona and women's wear in Madrid.

Well-known designers include Antonio Alvarado, Adolfo Dominguez, Purificación Garcia, Roser Marcé Manuel Marino (for the Roberto Verinno label), Marguerita Nuez, Jesús del Pozo, Nacho Ruiz, Sybilla, and Vitorio & Lucchino (actually José Victor Rodriguez Caro and José Luis Medina del Corral). Spain is also an important producer of leather products, particularly shoes, and also moderately priced knitwear.

JAPAN

Tokyo, the capital of Japan and the center of their fashion industry, enjoyed the position of major international trendsetter in the 1980s.

Until World War II, Japanese women still wore the kimono. As they started to take jobs in the postwar period, practical Western dress became a necessity. In the 1950s and 1960s the Japanese became eager for Western-style clothes, most notably those from Paris. Their fascination with the West continues as many large apparel firms have licenses to manufacture European and American fashions in Japan.

The "first generation" international Japanese designers, Hanae Mori and Jun Ashida, held shows in Paris in the 1960s. In the 1970s Kenzo Takada became an international success working out of Paris. Since then, other designers such as Junko Koshino, Mitsuhiro Matsuda, Issey (Kazanaru) Miyake, Rei Kawakubo (Commes des Garçons), Kansai Yamamoto, and Yohji Yamamoto have gained international recognition. Their creations have given full play to the Japanese aesthetic sense, which is rooted in a culture where Western garments were not traditionally worn. In the 1980s these designers influenced Western fashion with their unusual shapes, somber color combinations, asymmetrical balance, layering, interesting textures, and use of natural fibers, particularly cotton.

Most Japanese manufacturers are small or medium sized, but there are a few operating on a very large scale, including Renown, Onward Kashiyama, Sanyo Shokai, and World Company. These firms not only produce Japanese designs and license European and American designs, but also have been buying production facilities and apparel companies in the West.

CANADA

Canadian designers blend European and North American styling and export many of their products to the United States under a free trade agreement.

The apparel industry is the eighth largest industry in Canada, employing more than 80,000 people and accounting for an estimated 2000 apparel firms. The two largest centers are around Montreal, Quebec, with approximately 57 percent of the apparel workforce; Toronto, Ontario with some 31 percent; followed by Manitoba province with 6 percent.[7]

Well-known **Montreal** area designers and collections include Jean Airoldi, Helene Barbeau, Angela Bucaro, Simon Chang, Leo Chevalier, Denomme & Vincent, Michel Desjardins, Denis Desro, Margaret Godfrey, Robert Krief, Dita Martin, Jean Claude Poitras, Hilary Radley, Michel Robichaud, Marie Saint Pierre, and Debbie Shuchat.

Over 500 textile or fashion companies in Ontario employ 26,000 people. In **Toronto** itself, more than 14,000 people work for 300 fashion-related companies concentrated in the King-Spadina Fashion District, making fashion Toronto's largest industrial employer. An important new development is the Toronto Centre for the Promotion of Fashion Design. Known as the Fashion Incubator, it provides two years of manufacturing facilities at minimal costs for eight fortunate fledgling designers in order to get them started in a successful business. Toronto-area designers and collections include Lida Bada, Brian Bailey, Dominic Bellissimo, Marilyn Brooks, Stephan Caras, Tricia Cochrane, Judy Cornish and Joyce Gunhouse for Comrags, Roger Edwards, Edie Johne, Debora Kuchme, Paula Lishman, Linda Lundstrom, Mariola Mayer, Pat McDonagh, Franco Mirabelli, Price Roman, Anne Seally, Alfred Sung, Donna Stephens, Michael Tong, Nancy Young and Karen Gable for Zapata, and James Yunker.

Two of the largest Canadian manufacturers, Freed & Freed International (which owns JMJ

Canadian designer Jean Claude Poitras with his Fil D'Or (Golden Thread) award at the International Linen Festival. *(Courtesy of the International Linen Promotion Commission)*

Canadian designer Marilyn Brooks with her models at a collection opening.
(Courtesy of the Toronto Ready-to-Wear Collections)

Fashions, formerly Sterling Stahl) and Nygard International (Peter Nygard), are located in Winnipeg. Well-known companies in Vancouver include Mr. Jax (designed by Ron Leal), NETO leathers, and Virani.

Although quite diversified, Canada's apparel industry is particularly known for exports of men's tailored clothing, outerwear, furs, leather goods, and children's wear. The Fashion Apparel Collections Exhibition (FACE) shows in Montreal in February and in Toronto in August. Toronto also plays host to the Toronto Ready-to-Wear Collections in February and September. Permanent showrooms are found and buyers' weeks are held at the Ontario Fashion Exhibitors (OFE) in Toronto and the Place Bonaventure in Montreal. There is also a newly established Canadian Apparel Federation to promote the industry.

A free-trade agreement was made between the United Sates and Canada in 1989. Although it provides for eventual elimination of tariffs, it has a yarn-forward clause which stipulates that all yarns and fabrics must come from the U.S. or Canada. The problem for Canadian designers is that many of them have based their unique styling on the use of European fabrics. Since 1988, exports to the United States have doubled. Canada is the second largest (after Italy) exporter of men's suits to the U. S.

THE UNITED STATES

American designers and manufacturers understand the domestic market best and are the most successful in selling volume fashion in the United States.

New York

As one commentator observed, "Each of the world's great fashion centers has its distinct specialty. Paris glorifies design for design's sake, Milan emphasizes casual luxury, London aims to shock, and Tokyo, to mystify. By contrast, New York produces fashion for Everywoman—straightforward, sensible, businesslike, and increasingly stylish."[8] American fashion blossomed during the war years, when communications to Paris were cut off. Since then, an American style, especially for sportswear, is increasingly appreciated worldwide.

New York became the American garment center because supplies and skilled labor were concentrated there. At least two-thirds of American fashion manufacturing is still located in New York: designing, manufacturing headquarters, the major markets, and apparel showrooms.

The Seventh Avenue Garment District

Most fashion originates in the vicinity of Seventh Avenue on Manhattan Island in New York City. Seventh Avenue gives its name to the whole garment district, which is bounded by Thirty-fourth and Fortieth Streets and Sixth and Ninth Avenues. Crowded into this area are over 2500 manufacturers and contractors. In New York City there are an estimated 4500 factories located primarily in Chinatown, Manhattan and Sunset Park, Brooklyn. The garment industry employs approximately 95,000 people and is still the largest manufacturing industry in New York City. This represents approximately 8 percent of overall apparel employment in the United States.[9] In order to preserve manufacturing in the city, owners of these buildings are now blocked from converting more than half of their space from manufacturing to offices. Many buildings in the garment district are known for certain apparel specialties. For example, 550 Seventh Avenue has traditionally housed high-fashion companies, one on each floor.

Seventh Avenue is anything but glamorous. It is a conglomeration of old, run-down buildings. The streets are filled with trash; jammed with trucks and taxicabs; young men speaking foreign tongues push racks of fabrics or clothing from suppliers to manufacturers; sales representatives pull suitcases on wheels; and the homeless congregate on the sidewalks. Inside, design studios, showrooms, and offices crowd each building on every block. There is a sharp contrast between the plush showroom in front, for the outside world to see, and the cluttered design rooms in back. Yet this place is the fashion capital of the United States and one of the major fashion capitals of the world!

Company size and ownership

Traditionally fashion businesses have been small, often family owned and operated. Their small size allowed flexibility in both design and production, which is needed to respond quickly to market needs. Today, however, the sizable advertising and marketing budgets of the large companies make it difficult for small companies to compete. As a result, small companies have been bought up by larger ones so that the number of apparel firms has decreased. However, overall production has increased with the growth of the larger companies. Liz Claiborne, the world's largest women's apparel company, has many divisions including Lizwear, Lizsport, Elizabeth, Liz Claiborne Petites, Dana Buchman, Claiborne men's wear, and accessories.

American fashion companies also reflect globalization. Many foreign companies have invested in the United States, purchasing companies or production facilities. Takihyo, a Japanese firm, backed Donna Karan so that she could start her own business. The industry is full of similar examples.

Designers

In companies that produce moderate or lower-priced clothing, the designer's name is usually unknown to the public. The company may use a fictitious name such as "Leslie Fay." With high-priced clothing it is a different matter. Designers who have proven themselves may have their names added to the company's label such as "Richard Tyler for Anne Klein". Others start their own businesses under their own names. If their collections are successful, if they have skillful business and financial partners, quality production, and clever advertising, their names become well known. As in show business, however, designers are only as good as their last production. A designer is often a star today and forgotten tomorrow; the picture changes every season. It is increasingly difficult to name the most important American designers because they change from year to year. Reading fashion publications regularly is the only way to keep informed about current designer favorites and best-selling styles.

Well-known American designers

Geoffrey Beene. Born in Louisiana in 1927, Beene is a consistently innovative designer. He originally studied to be a doctor but complained that "every disease we studied I got." He began his career in the display department of I. Magnin in Los Angeles. In

American designer Donna Karan with her models at a collection opening. *(Courtesy of The Donna Karan Company, © DK Co.)*

the 1940s he studied fashion in New York and with Molyneux in Paris. He designed for Teal Traina from 1958 until he started his own collection in 1963 with partner Leo Orlandi. He later added men's wear and Beene Bag sportswear. One of the first American designers to show in Europe (in 1975), he has over 30 licenses.

Bill Blass. Born in Indiana in 1922, Blass worked as a sketcher for David Crystal in the 1940s. He worked as designer for Maurice Rentner, later became a partner, and then in 1968, bought out the business. To his women's day and evening collections he has added Blassport sportswear and men's wear. With 30 domestic and 70 international licenses, Blass has had the longest continuing success of any American designer.

Louis Dell'Olio. Born in New York City in 1948, Dell'Olio attended Parsons School of Design with Donna Karan. After spending a summer working for Norman Norell, he became an assistant at Teal Traina and later sportswear designer for Originals. He joined Anne Klein in 1974, as codesigner until 1985 and then sole designer until 1993.

Donna Karan. Born in 1948 in New York to parents in the "rag trade," Donna (Faske) Karan left Parsons to become Anne Klein's assistant. After Anne Klein's death in 1974, Karan became head designer with Louis Dell'Olio. This was the first time that an American fashion company was able to continue successfully without the original designer. In 1985 Tomio Taki offered to back Karan in her own business featuring luxury sportswear. Her bridge collection called DKNY (Donna Karan New York) has been extremely successful.

Calvin Klein. Born in New York City in 1942, Klein always wanted to design clothes. He studied at F.I.T. (Fashion Institute of Technology) and worked at Millstein sport and suit company on Seventh Avenue. In 1968, he and his friend Barry Schwartz opened a coat business and got their first order when a buyer from Bonwit Teller accidentally got off the elevator on the wrong floor. He expanded into sportswear, men's wear, jeans, and accessories. He was the first American designer to open his own shops in London and Milan. His designs consistently represent the clean all-American look.

Ralph Lauren. The perfect example of a designer who was able to build an empire on a life style, Lauren was born in New York City in 1940 and began his career as a salesclerk at Brooks Brothers. In 1967 he started to design ties and by 1968 had established Polo men's wear with backing from Norman Hilton. The Polo name is perfect for his classic Ivy League

American designer Ralph Lauren at a collection opening. *(Courtesy of Ralph Lauren)*

look in expensive fabrics. Lauren built the rest of his multimillion-dollar business on licenses for women's wear, a less expensive men's wear collection called Chaps, boys' wear, girls' wear, accessories, and home furnishings. In 1986 he opened a $14 million retail store on Madison Avenue and has Polo shops in stores across the country.

Nicole Miller. Born in Texas in 1951, but raised in Massachusetts, Miller studied at the Rhode Island School of Design and in Paris. In 1982 she opened her women's wear collection with partner Bud Konheim. What began as a way of getting rid of excess fabric turned into a huge additional business of whimsical men's ties, shirts, and boxer shorts in witty prints. She now has Nicole Miller boutiques in Mexico, Canada, Argentina, Spain, and Korea.

Isaac Mizrahi. Born in Brooklyn, New York, on October 14, 1961, Mizrahi studied at Parsons and apprenticed with Perry Ellis. He opened his own women's wear business in 1987 with partner Sarah Haddad-Cheney and added a men's wear collection in 1990. Mizrahi says, "With each collection, I try to reach the tap roots of what is truly American—clothes that are pared down, comfortable and practical, yet have a sense of luxury and an element of wit and surprise."[10] Mizrahi also likes to design costumes for the ballet and films. In 1992 he won his third Council of Fashion Designers of America (CFDA) Designer of the Year Award.

Oscar de la Renta. Born in 1933 in the Dominican Republic, de la Renta studied painting, sketched for Balenciaga in Madrid, and afterwards became assistant to Castillo at Lanvin. In 1962 he designed for Elizabeth Arden in New York and in 1965 became partner at Jane Derby where he took over the business in 1966. His work, which enjoys a reputation for elegance, includes evening wear, suits and dresses, sportswear, men's wear, accessories, and the less expensive Miss O collection. Now he also designs couture for Balmain in Paris, the second American to do couture in Paris.

Arnold Scassi. Born in Montreal, Canada, Scassi (his family name *Issacs* spelled backwards), studied art and later fashion design in Paris. After working for the French house of Pacquin and the American designer Charles James, Scassi opened a ready-to-wear business in the late 1950s. In 1964 he gave that up to concentrate on two couture collections a year. In 1984, he again added ready-to-wear including day and evening dresses, bridal, sleepwear, and handbags. He recently received the CFDA Award for evening wear.

Richard Tyler. A native of Australia and based in Los Angeles, Tyler first developed his design reputation with men's wear and then added women's wear. He is well known for his meticulous tailoring and Hollywood clientele. After a successful first showing of his collection in New York, he was suddenly projected into the fashion limelight when he was chosen as design director for Anne Klein and Co. in New York. He continues his signature line in partnership with his wife, Lisa Trafficante.

Adrienne Vittadini. Born in Hungary, Vittadini moved to the United States as a young girl and studied fashion in Philadelphia. After gaining experience at Louis Feraud in Paris, she designed knits for Rosanna and Kimberly in the United States. In 1979 she struck out on her own with women's knit sportswear, activewear, and more recently dresses. She is involved with the design of her own fabrics as well as apparel.

Other successful designers with signature collections include Joseph Abboud, Adolfo, Linda Allard for Ellen Tracy, Randolph Duke, Mark

Eisen, Gordon Henderson, Carolina Herrera, Marc Jacobs, Andrea Jovine, Gemma Kahng, Norma Kamali, Randy Kemper, Michael Kors, Byron Lars, Bob Mackie, Mary McFadden, Rebecca Moses, Josie Natori, Todd Oldham, Carmelo Pomodoro, Mary Ann Restivo, Christian Francis Roth, Gloria Sachs, Ronaldus Shamask, Anna Sui, Zang Toi and Joan Vass. Men's wear designers include Joseph Abboud, Tommy Hilfiger, and Ralph Lauren.

Regional Fashion Centers

The United States has approximately 22,000 apparel manufacturers employing over a million people.[11] Although New York City remains the largest fashion center, the trend toward decentralization is increasing. Some manufacturers have left New York City in search of a cheaper labor supply and more space. Other companies have started up in regional locations. The result is the growth of regional fashion centers.

California

California is the second largest apparel manufacturing state, with about 5000 manufacturers and contractors and one-fourth of the nation's apparel production. The California clothing industry has a reputation for innovative styling and the advantage of close proximity to sourcing in Asia.

Much of the California designers' success is based on their ability to promote a certain life-style approach to fashion. California is especially known for sportswear, swimwear, and contemporary dresses. Sales of the *California* look keep growing as Americans become more leisure conscious and as population shifts toward the sunbelt.

Los Angeles. At least two-thirds of California's fashion companies are headquartered in Los Angeles County, making it the second largest U.S. apparel center. Nationally known designers and fashion brands in Los Angeles include Katayone Adeli for Parallel, Christine Albers, Barbara Barbara, Patti Cappalli, Rosemary Brantley, Eletra Casadei, Christian de Castelneau, Cherokee, Cross Colours, David Dart, Luis Estevez, Dennis Goldsmith, Tina Hagen, Karen Kane, Mara Heinz, Nancy Heller, Carole Little, Georges Marciano for Guess, Jonathan Martin, Leon Max, Robin Piccone for Body Glove, Allen Schwartz for ABS, Shelli Segal for Laundry, Bonnie Strauss, James Tarantino, and Sue Wong.

San Francisco. The nation's third largest fashion city, San Francisco is home to approximately 400 apparel companies. The garment industry is San Francisco's second biggest industry and its biggest source of manufacturing jobs. It is the headquarters for companies such as Levi Strauss, Esprit, the Gap and its subsidiary Banana Republic, Byer, Koret, and Lilli Ann. Local designers include Michael Casey, Isda Funari, Jessica McClintock, Lat Naylor, and Celia Tejada.

Other Regional Centers

Other important apparel manufacturing and design centers are scattered throughout the country.

Seattle area companies such as Roffe, International News, Olympic West Sportswear, Generra, and Union Bay have made it an important men's sportswear and outdoor/skiwear center.

Dallas is home to designers and companies such as Howard Wolf, as well as Susan Apple, Victor Costa, Michael Faircloth, Haggar, Jo Hardin, Rebecca Harrison, Holly Holman, Jerrel, Ginnie Johansen, Megan Moore, Ann Tobias, and Howard Wolf.

Chicago (the home of Hartmarx), Rochester (Hickey-Freeman), and Philadelphia (Botany 500 and After Six) have long been known as centers for men's wear. Philadelphia (Albert Nipon and J. G. Hook), St. Louis, Miami, and Boston also have some apparel and accessory manufacturing. As regional manufacturing develops, each center becomes less and less specialized. The spreading of fashion centers throughout the United States and the world is helping to balance fashion influence and to diversify styling.

SUMMARY

Fashion has become a global phenomenon. Due to a concentration of resources, supplies, skilled labor, and creativity, Paris grew to be the fashion capital of the world. It built its reputation with the couture but profits now come from prêt-à-porter. Today, Paris shares the European spotlight with Milan whose designers and fashion brands have achieved similar international success. London was the focus of youthful fashion of the 60s and remains noteworthy for its Savile Row tailoring. Also in Europe, German and Spanish designers are gaining global respect. In Asia, Tokyo designers had great impact on fashion in the early 80s. Canada has its style centers in Montreal and Toronto. New York, especially the area around Seventh Avenue, is the fashion capital of the United States and continually gains international recognition. Los Angeles and San Francisco make California the second largest U.S. center. Fashion has become big business and the top designers are the fashion stars.

CHAPTER REVIEW

Terms and Concepts

Briefly identify and discuss the following terms and concepts:

1. Paris as a fashion capital
2. Couture
3. Prêt-à-porter
4. Atelier
5. Toile
6. Salon
7. Milan as a fashion center
8. Savile Row
9. London as a fashion center
10. Tokyo as a fashion center
11. Seventh Avenue

Questions for Review

1. Why does Paris continue to remain a fashion capital?
2. What are the requirements for membership in the Chambre Syndicale de la Couture?
3. Explain the two classifications of ateliers and discuss the organization of a typical atelier.
4. What are the differences between the couture and the prêt-à-porter?
5. Discuss decentralization of fashion centers in Germany and Spain versus centralization in England and France.
6. What are the fashion centers in Canada? How are Canadian designers affected by the trade agreement with the United States?
7. What name is used to refer to New York's garment district? Explain why that name is used.
8. Discuss the growth of regional fashion centers in the United States.

Projects for Additional Learning

1. In a fashion magazine, find a photograph of a garment you particularly like. Find the name of the designer in the description. Trace that designer's name through older issues of fashion magazines until you have found ten examples (in ten separate issues) of his or her work. Analyze the unique characteristics of the designer's style.
2. Shop a store that carries Italian or French fashions. Find five of the names mentioned in this chapter on garment labels. Discuss the style characteristics of each garment in a written report. If possible, illustrate with sketches. If a store is unavailable, find your examples in magazines.
3. Shop an exclusive specialty store and ask to see their most expensive evening dress or suit ($1000 and up). Then go to a department store and examine an evening dress or suit in the $400-$600 range. Finally, go to a discount store and look at an evening dress or suit in the $100-$200 range. Make a written comparison of the quality of fabric, construction, and styling. Document your report with descriptions or sketches.

NOTES

[1] Barbara Weiser, owner of Charivari, as quoted in "Great American Designers, Do They Really Matter?" *Women's Wear Daily,* March 29, 1993, p. 6.

[2] As quoted in "The Concorde Couturier," *Women's Wear Daily,* November 17, 1992, p. 24.

[3] As quoted in "Couture Rule Changes Lead," *Women's Wear Daily,* October 21, 1992, p. 33.

[4] Bernadine Morris, "After Paris Shows, Haute Couture Finds Itself at Odds Over Function," *The New York Times,* February 2, 1993, page B4.

[5] As quoted by Marylou Luther, the Fashion Group International Trend Report, August, 1992.

[6] Heidi Lender, "The Latest From Lagerfeld," *Women's Wear Daily,* February 9, 1993, p.4.

[7] Audrey Gostlin, "Inside Fashion" newsletter, January 26, 1993, p. 5.

[8] Perdrix, "The Making of New York Fashion," *Connoisseur,* February 1986, p. 64.

[9] Arthur Friedman, "Staying Alive," *Women's Wear Daily,* August 25, 1992, p. 12.

[10] As quoted in Mizrahi press biography, 1993.

[11] Joanna Ramey, "8000 Apparel Industry Jobs Lost in August," *Women's Wear Daily,* October 5, 1992, p. 19.

Parisian designer Christian Lacroix with sketches of his collection.
(Courtesy of Christian Lacroix)

9

PRODUCT AND DESIGN DEVELOPMENT

CAREER FOCUS

Responsibilities for product development, design, and merchandising vary from manufacturer to manufacturer. When the designer is the president of the company, management of product development and design is less complicated: the designer is obviously in charge. In other companies, product development is usually a joint effort between management, designers and merchandisers. Manufacturers that are design-driven, like Liz Claiborne, have designers in charge of product development with merchandisers working for them. In companies that are merchandising-driven, like Levi Strauss, merchandisers or product managers direct development of the line and design. In a large company, a designer, a merchandiser and their assistants are assigned to each product group. There are also pattern and sample makers involved in sample production.

163

CHAPTER OBJECTIVES

After reading this chapter you should have attained competence in the following areas:
1. Understanding line development by item or by group.
2. The ability to explain the important elements and principles of design and their application to line development.
3. The ability to describe the process of creating a sample garment.

The three major divisions of a clothing manufacturer are *design*, *production*, and *sales*. Each is related to and dependent upon the others. The design department creates the new styles within the company's image or identity; the sales department markets each line, acting as a liaison between the manufacturer and the retailer; and the production department produces the line in all required sizes and fabrics, filling retail store orders. Financial management is also very important. This chapter covers product development, Chapter 10 explains production, and Chapter 12 explores marketing and sales. This chapter begins by explaining how a manufacturer's line is developed. To prepare for this, it is very important to first read Chapters 2, 3, and 4 to understand design influences and resources. The chapter then goes on to discuss fashion design elements and principles, including color and fabrication, and the creation of the sample line.

PRODUCT DEVELOPMENT

Management, merchandisers, designers, and their assistants are involved in the development of a line or collection of the fashion manufacturer's product.

Appealing to a Target Market

Each apparel manufacturer is defined by its customer and identified by its particular style. Manufacturing is said to be *market driven* when it responds to the needs of consumers. Designers and manufacturers must identify their customers as a group, develop a product that fits that identity, and stay with it. This involves finding a *market niche* around a particular life

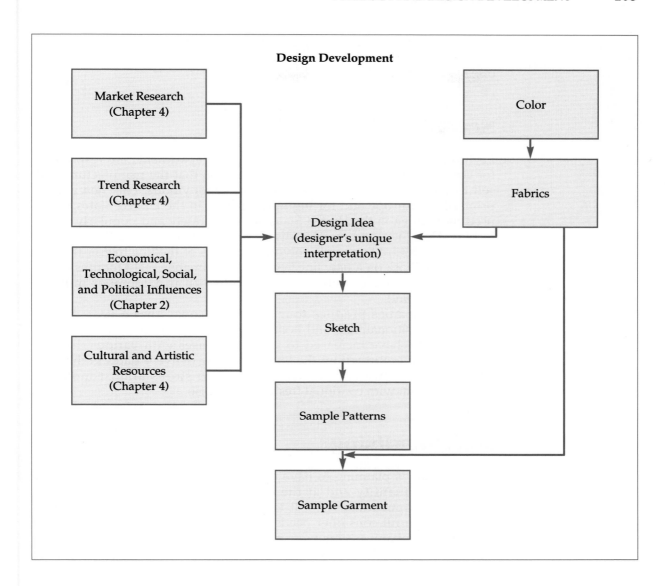

Design Development

Market Research (Chapter 4)

Trend Research (Chapter 4)

Economical, Technological, Social, and Political Influences (Chapter 2)

Cultural and Artistic Resources (Chapter 4)

Color

Fabrics

Design Idea (designer's unique interpretation)

Sketch

Sample Patterns

Sample Garment

style or need. Christian Lacroix reflects on his own creative methods: "As I look for my line, I ask myself what will be the life style of the woman of my next collection."[1] A designer or manufacturer that develops and is recognized for a particular style, yet incorporates current trends into that style, attracts buyers year after year. Lacroix, Ralph Lauren, and Jessica McClintock are excellent examples of designers with a distinctive style. Lauren himself said, "It's the most important thing for designers...to have an identity...."[2]

Traditionally manufacturers specialize in a particular styling category, price range, and size range of apparel. As a company grows, it expands by adding diversified lines. Many manufacturers have broadened their product lines to include other style categories or size and price ranges, but they have separate divisions and/or label names for each. For example, sportswear firms such as Liz Claiborne and Carole Little have added dresses. As manufacturers diversify, however, they must maintain a consistent identity for their customers. Giorgio Armani reiterates, "Coherence is always the secret, I believe, behind a great company."[3] Bob

Abajian, head of design at Liz Claiborne, adds, "It is important for a company to have a single design focus for the entire line and everything else to support that."[4]

Seasons

Each season, the design and merchandising departments are responsible for creating a new line, the seasonal collection that the manufacturer will sell to retail store buyers. The terms are synonymous: *collection* is used primarily in Europe and for high-priced lines in the United States; *line* is used more often in the United States for moderately and popularly priced fashion. Designers and merchandisers work on two or more lines at once, designing a future line while solving problems of the line that is about to be shipped. Work on a new line begins approximately eight months before the selling season (a velvet dress to be worn in December must be designed in May).

Most companies produce four or five seasonal lines a year: spring, summer, transitional, fall, and holiday or resort. Some companies such as Liz Claiborne have increased to six line releases with various delivery dates for each group within the line. Managing delivery dates in this way means that manufacturers are shipping to stores almost monthly, providing them with continual fresh merchandise.

Merchandising

Merchandising is planning to have the right merchandise at the right time in the right quantity and at the right price to meet the needs of the company's target consumers. It is the merchandiser's responsibility to make plans to fill consumer needs. The merchandiser's activities, which vary from company to company, may include making the merchandising plans, planning fabric purchases, scheduling production, and working with designers.

The dollar merchandise plan

Each season, the design department has to decide how many groups are needed to meet both the demands of retailers and the financial goals of the manufacturer. A *dollar merchandise plan* projects seasonal sales for each line. The previous year's actual sales are used as a basis for projected sales goals. Spread sheets are created to show what needs to be produced and sold per month to reach sales and profit goals. Computers have made this job much easier.

The merchandising plan

Next, the merchandiser develops a merchandising plan to determine the line concept and the approximate number of styles and fabrics required to meet sales goals. Price points are established so that fabric and labor cost limits can be determined. In the value-oriented 90s, manufacturers have tried to lower prices by offering less expensive fabrics or by cutting labor costs. The merchandiser sets up a schedule of deadlines for styling, finished samples, and production, working backward from the required shipping dates.

DESIGN DEVELOPMENT

The designer creates the styles for the collection, giving form to fashion ideas.

The emphasis on uniqueness has made creativity in design more important than ever. Couture and top ready-to-wear designers are able to enhance their creativity with expensive fabrications and beautiful workmanship because they are able to sell at higher prices. Their skill enables them to create trends. Mainstream designers must know how to adapt trends to suit their own customers and select fabrics that fit into their price range. Designers must also supervise pattern making and sample making and see their lines through to a successful completion. As British designer Jasper Conran remarked, "A collection isn't sitting down and doing sixty drawings. There's the commercial side and there's the poetic. The two have to marry. If you have only the poetic, then however fabulous it is you're going to be bankrupt."[5]

To begin work on a collection, many designers use *concept boards* to show their ideas to the management team. They make a collage of color and fabric swatches, sketches of ideas, and *swipes* (idea photos) from magazines that capture a mood or theme. When the ideas are approved, designers refine and develop specific ideas into themes for groups and individual garment designs.

Groups

Usually lines are divided into groups of garments. Liz Claiborne, for example, does 6 to 8 groups in each line release in each division. Each group has a specific *theme* based on a fabric, color, or a particular fashion direction. Ideas for the theme come from research and environmental influences. Sometimes the idea for a single garment may inspire an entire group. The styling of each garment within each group has variety yet carries out the central theme. To present a visually pleasing group of dresses, gowns, suits, or coats, a few common elements are needed, such as a fabric or color story.

Dresses

Groups of dresses may begin with an *anchor,* a best seller from the previous line with a change of color and/or fabric. Often dress designers emphasize only a few silhouettes, called *bodies,* interpreting each of them in several prints; or they will feature one print in a variety of styles. Within the group, the garments must offer a variety of silhouettes, sleeve treatments, necklines, and/or other details.

A group of sportswear from Esprit's Ecollection. *(Courtesy of Esprit)*

Sportswear

Sportswear differs from dress design in that individual pieces are designed to be worn together. A balanced mix of skirts, pants, jackets, and shirts or blouses and other tops are included in each group. There may be a basic jacket and a fashion forward jacket, various skirt lengths, a basic pant and an updated version, shirts plus other kinds of tops with a variety of necklines.

The number of pieces or styles in a group is determined by the statement the designer wishes to make and the variety of pieces needed to complement one another. The designer must think through what the customer can wear with what. A top has to go with more than one jacket and/or bottom. "Singular usage garments do not make it in the 90s," states Bob Abajian.[6] A *base fabric* is selected for jackets and lowers. Many companies build a reputation around one consistent base fabrication. In jeanswear, of course, the base fabric is always denim. There must be an assortment of styles and fabrics (solids and prints) to go with the base fabric to please various consumer tastes. At the same time there should still be consistency of theme, with color, fabric, line, or detail tying the whole group together.

Coordinated sportswear is designed to mix and match interchangeably as opposed to *separates* which do not show an obvious relationship. Coordinated sportswear is sold as a package to the retailer while separates are sold individually.

Men's sportswear is also merchandised in groups. Generally, a men's sportswear group consists of related separates, pieces that go together but without obvious coordination. A group may include pants, jackets, vests, shirts, and sweaters. The group still has a central theme carried out in a variety of related fabrics. Men's sportswear now represents one third of the men's wear industry.

Suits and coats

Unlike sportswear, suits and coats are treated as individual units sharing only a basic silhouette. The manufacturer's objective is to include a variety of colors and suitings in the line, providing a wide choice for both the retail buyer and the consumer. The prime factors in styling men's suits are fabric and silhouette. There is generally little detail change from season to season. The ideal suit is a timeless classic. Women's suits and coats, on the other hand, may be fashion oriented or classic. Like dresses, a group must offer a variety of silhouettes, sleeve treatments, necklines or other details.

Fashion versus basics

Every group must have a balance of fashion and basics. Even if the manufacturer is not a designer business, fashion pieces give the company credibility by showing retail buyers that the company is aware of fashion trends. Basics are often what brings in the money to allow them to do the fashion. Even designer lines such as Ralph Lauren make most of their sales and profits on one or two hot basics such as khaki pants or polo shirts. The strength of basics is also obvious at Levi Strauss. Although they offer many varieties of fits and colors in jeanswear, their biggest seller is still the 501 basic jean.

Items

Some manufacturers produce single items that do not relate to each other. *Items* are popular style categories that a consumer can use to update an existing wardrobe. Each garment must be strong enough to sell on its own, not depending on the strength of other garments in the group. Such garments are usually produced at moderate or budget price ranges at the request of retailers for private labels (see Chapters 12 and 13) and/or by a knockoff house.

Knockoffs

A *knockoff* is a copy of someone else's design, usually a garment that is already a best-seller for another manufacturer. For that reason, producers that copy can safely invest in volume production. They simply buy a particular garment, make a pattern from it, order large quantities of the same or similar fabric, and have the garment manufactured. Production and fabric costs are lower because of the huge quantities produced. Moreover, the knockoff house does not have to spend money on design development. The knockoff producer must have (a) an acute awareness of what garments are selling well at the retail level, (b) rapid production capabilities to capitalize on the success of the style while it lasts, and (c) lower prices.

Some designers try to copyright their designs. Copyright is nearly impossible, however, because of the fast pace of the industry. By the time a design is granted a copyright, the garment is no longer in fashion. To protect their profits, some designers copy their own designs at lower prices. When Richard Nott and Graham Fraser at Workers for Freedom realized that they were being copied, they started a lower- priced collection (in lower-priced fabrics) to copy their own line .

Copying becomes a more serious problem when it involves regarding counterfeit merchandise (imitations with fake labels intended to deceive the consumer). Fake Gucci and Ferragamo accessories are seen all over the world. The U.S. government confiscates this merchandise when they discover it. It is difficult to control counterfeiting internationally, however, because a designer or brand name must be registered separately in every marketplace. The cost of registration plus legal fees is very high when it must be paid in every country.

Design Elements

Keeping the theme of the group in mind, a designer must incorporate into each garment a pleasing combination of all the elements of good design: color, fabric, line, and shape. The ingredients of design, which are essential to every art form, are not a recipe for success and cannot substitute for experience. The designer does not think of these elements consciously, they become a habit. Fashion changes continually; there are no hard and fast rules.

Color

Color is the first element to which consumers respond, often selecting or rejecting a garment because of its color appeal. Therefore, designers must consider their customers and provide colors that are both appealing and flattering.

People connect certain colors with holidays and seasons. They expect to see earth tones in fall clothing, jewel colors for the holidays, the pastels of flowers in the early spring, and refreshing white for the summer. All manufacturers include some of these colors in their lines.

Color dimensions
Color has three different dimensions: hue, value, and intensity.

Hue enables us to tell one color from another, such as red from blue or green.

Value refers to the use of darks and lights; it is the variation of light strength in a color. The value scale runs from white to black. White is pure light, black is the total absence of light; adding white lightens a color and adding black darkens it. The lighter values are called *tints* and the darker ones *shades*. Every garment has value contrasts, even if only those created by normal gathers and folds. Strong value contrasts (pure black against pure white is the strongest) achieve a dramatic effect.

Intensity is the relative brightness (strength) or paleness (weakness) of a color. Bright colors are considered high intensity, pale ones are low intensity. For instance, when paint is paled by adding water, the intensity of its color is lowered. Within the blue hue, marine blue is high intensity and soft pastel blue is low intensity.

Warm colors
We classify red, yellow, and orange as warm colors because of their association with fire and the sun. Warm colors are stimulating, aggressive,

and lively. Red is associated with matters of the heart: valentines, love, and romance; it is also exciting, fiery, and dangerous. A popular color for sportswear and evening wear, it is one of the few colors used in high intensities for clothing in every season. Yellow is bright, sunny, cheerful, friendly, and optimistic but it can be a difficult color to wear because it conflicts with many skin tones. Orange combines the sunniness of yellow and the warmth of red. It is often a difficult color to wear as intense orange can be irritating and overpowering. Usually it is best toned to a softer peach or apricot or limited to use in active wear.

Cool colors

Cool colors, as the term implies, are refreshing in contrast to warm colors. The cool colors—blue, green, and purple—remind us of the sky and the sea. Blue is quiet, restful, and reserved. Denim blue and navy have become wardrobe classics. For that reason most manufacturers include it in at least their spring or summer line. Green is a refreshing color, suggesting peace, rest, calm, and quiet. It is used primarily in a dark value in fall sportswear lines or mixed with neutrals to create earthy olive or loden. Purple, historically associated with royalty, has come to represent wealth, dignity, and drama.

Neutrals

For sophisticated fashion, neutrals such as beige, tan, taupe, brown, white, gray, and black are even more popular than the colors just mentioned. The reason is probably that they present a pleasing background for the wearer without competing for attention. Neutrals are part of every season's fashion picture, as either a strong fashion statement or a way to round out a color story.

White is associated with purity and cleanliness. Because it reflects light, it is cool in the summer. In Western culture black has been connected with villains and death. However, it has come to be regarded as a sophisticated fashion color. Both black and white have become classics and therefore tend to stay in fashion longer.

An imaginative and unusual combination of colors in a Christian Lacroix design. Also note the successful use of all elements and principles of design. *(Courtesy of Christian Lacroix)*

Color relationships

There are no hard and fast rules for the use of colors. Rather, colors are considered harmonious if they are used so that one color enhances the beauty of the other. Colors are now combined in many more unusual ways than ever before. Ethnic influences on fashion have changed our view of color combinations, making us more receptive to new ideas. In addition, colors run in fashion cycles just as styles do. Color combinations that look right to us now will not work with the new looks a few years hence.

Color naming

An exciting color name can be important in promoting a fashion look. Fashion colors sometimes reappear with new names that make them seem fresh. "Plum" of one year might return as "aubergine" in another. Oil paint color lists, as well as books featuring the names of flowers, trees, wood, fruit, vegetables, spices, wines, gems, and animals, can provide color-name ideas. To create moods, colorists and fashion journalists use exotic names such as "China Blue" or "Poison Green."

Color selection

Colors must be selected for each group and for individual garments. Within the line, each group is usually formulated around a color plan consisting of as few as two colors or many colors. The colors in a print fabric may inspire a color story.

Color choices must reflect season, climate, and type of garment. Active sportswear, for example, employs many more vivid colors than business attire. Every line should include a range of colors that appeals to a variety of customers.

As trendsetters, couture and top ready-to-wear designers have the privilege of basing their color story on whatever inspires them. Mainstream fashion designers may rely on color selections provided by color and design services, trade associations, or fiber companies. The colors common to these sources indicate trends. By basing a color story on these trend forecasts, the designer is assured that the group will be in the mainstream of fashion. Usually designers select some trend colors and combine them with their own color choices. The group may then be anchored with neutrals, darks, white, or black. In any case, the color story must be meaningful, not just a group of unrelated colors. It might be all brights, or all muted, or a balance of darks and lights.

Fabric

Color is interpreted in the medium of fabric. Fabrics are the designer's artistic medium, in that fashion design is essentially sculpture in fabric in relation to the body.

Fabrication is the selection or creation of an appropriate style for a fabric—or the reverse, the selection of the right fabric for a design. Fabrics themselves often inspire garment design. For example, the softness and drapability of a jersey might inspire gathers in a dress. Christian Dior wrote, "Many a dress of mine is born of the fabric alone."[7] Other designers work the other way around, first getting an idea, perhaps developing it in a sketch, and then finding the appropriate fabric for it. An unexpected use of fabric can create a fresh look, for example, a baseball jacket made from silk taffeta for evening wear. Many firms build a line or even an entire reputation on one fabric such as denim or stone-washed silk.

However the designer works, he or she must ultimately decide which fabric will work best with a design. Designers must develop the ability to picture a design already made up in the fabric. This ability comes through observation and experience. Next to understanding the needs of the

Robert Abajian, design director for Liz Claiborne, has a meeting with Jean Benz and Patricia Pastor to select fabrics for the knitwear collection. *(Photograph by the author)*

customer, choosing a fabric suitable for a particular style is probably the most important aspect of designing. The designer chooses fabrics on the basis of fashion trends, quality, performance, price, and suitability. The designer or merchandiser must be sure to include a variety of weights, textures, and prints in a line, as well as a balance of fashion and classic fabrics.

Fabric characteristics

Fabric suitability is determined by characteristics such as fiber, weave, texture, performance, hand, pattern, and color.

Texture is the sensuous element of design. It is the surface interest of a fabric, created by the weave and by light reflection. Our eyes appreciate the play of light on smooth or rough surfaces; we feel the surface with our hands; and sometimes we can even hear the texture, such as the rustle of taffeta. Combinations of textures, such as suede with jersey (rough with smooth), create interest in a garment.

Performance refers to a fabric's wearing and cleaning properties based on fiber content, weave, and finish.

Weight and *hand* dictate the silhouette of a garment. *Weight* is the heaviness or lightness, thinness or thickness of a fabric. *Hand* is the feel, body, and fall of a fabric. A designer must know how a fabric will behave and whether it will carry out an idea. A garment must be styled in a fabric that is compatible with the desired silhouette.

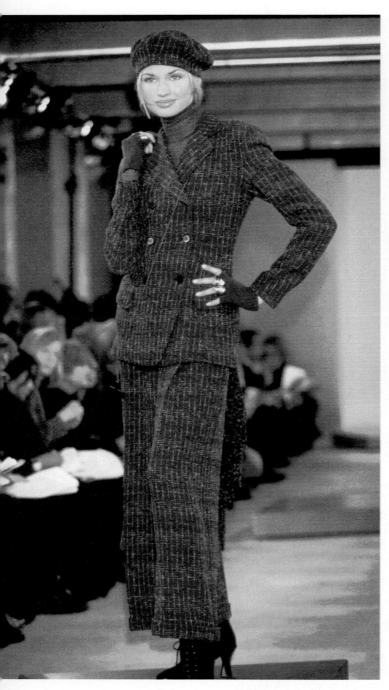

Fabric texture adds interest to a classic pantsuit by Donna Karan. *(Courtesy of The Donna Karan Company ©DK Co.)*

Firm fabrics like worsted wool, gabardine, and linen are needed to carry out a tailored look. These fabrics have both the crisp look and the fall necessary to achieve the desired effect. Interfacing, a fabric sewn or ironed into the garment's inner construction, is used to give additional stiffness to finished edges such as necklines, collars, cuffs, and buttoning areas.

Soft fabrics such as crepe, jersey, chiffon, and challis are ideal for draped designs that delineate body shape. Additional fluidity can be achieved by cutting fabric on the bias grain, so that the diagonal of the fabric falls vertically in the garment. However, bias is difficult to lay out on a marker (pattern layout), and thus expensive to produce. Softness may be increased by the additional use of gathers, shirring, smocking, and unpressed pleats.

Fabric weight varies with the type of garment. For example, blouse weights are lighter than bottom (skirt or pant) weights. Manufacturers often buy fabrics in specific weights, such as a five-ounce shirting or an eleven-ounce denim. The weight is determined by the square yard for wovens and by the linear yard for knits.

Fabric weight must also be appropriate for the season. Heavier, warmer fabrics are needed for winter and light, cooler ones for summer. As we saw in Chapter 5, specific natural fibers have traditionally been considered appropriate for certain seasons. Wool is used for fall and winter because it is heavy, bulky, and warm. Linen and cotton are used as warm-weather fabrics because they are light, cool, and washable. However, there are now seasonless fabrics and fashions. A fabric such as jersey can be worn year-round. Cotton, traditionally a summer fiber, can be woven into warm, bulky fabrics such as corduroy for fall and winter garments. Wools can be woven into lightweight voile for spring.

Prints and patterns

The *scale* of a pattern must complement the design. To show a large print to its best advantage, it should not be cut up with seams and details, but rather allowed to be the most important element of the garment; construction should be kept simple. On the other hand, if a design idea begins with dominating lines and details, then the fabric must be of secondary importance.

The *repeat* of a print is the amount of fabric necessary for a floral or geometric pattern to duplicate itself totally. Large repeats are not usually suitable for trims or for children's wear. In addition, garments to be made in border prints must be carefully thought out so that fabric is not wasted in production cutting.

The designer must recognize patterns that require special matching because they increase the amount of fabric used and therefore the cost. Bold plaids and irregular stripes must match when sewn together.

One-way prints and pile fabrics also use more yardage, because all the pattern pieces must be cut in the same direction. Most prints are two-way (that is, they have a motif that faces both directions), which makes cutting simpler and more economical.

Environmental concerns

Many designers are concerned with the condition of our polluted environment and want to do their part in trying to improve the situation. Companies such as Patagonia, Ecotex, Ecosport or Esprit with their Ecollection are trying to fill the need for environmentally responsible apparel. They feel that it is the fashion industry's responsibility to make environmentalism fashionable because fashion is an important communicator of values to people. To do this they must choose organic cotton, naturally colored cotton, linen, wool, or tencel which are treated with low-impact or bifunctional dyes, enzyme washes and other safe finishes. Some companies are returning to natural materials for buttons and using nonelectroplated metal trimmings. Patagonia is offering a jacket made out of recycled plastic!

Reviewing fabrics

The designer's involvement in fabric selection for a manufacturer varies. Sometimes designers have the entire responsibility for fabric selection; sometimes they work with a fabric merchandiser, especially when an entire group is built around a particular fabric.

Before a new season, a designer develops a feeling for fabric trends by reviewing the fabric market. The designer may visit one of the international fabric trade fairs or fabric libraries for an overall picture of the season's textile offerings. Designers and merchandisers frequently travel to a major fashion capital such as New York, where most textile mills and converters have headquarters. This is the best way to keep abreast of the newest fiber or fabric developments that might be the source of ideas for garments. Moreover, at these centers the designers can discuss their ideas directly with textile designers and technicians. Designers often work directly with mills or converters to develop a new pattern or fabric. Fabric mills or converter sales representatives also call on the designer at work. Each season designers try to see as many fabric representatives as possible to learn about the variety of fabrics available.

Price considerations

Fabric quality, and therefore price, must be consistent with the price of the line. An expensive designer collection is made of the finest fabrics; a moderately priced line requires less expensive materials. Fabric quality must compare favorably with competing lines. A general rule is that fabric and trimming costs must balance labor costs. More expensive fabrics may go into understated, simple garments whose construction is less complicated. Garments requiring more yardage have to be cut from less expensive goods, so that they will not exceed the price range of the line. By using a less expensive fabric, the designer can afford to put more money into labor or trimmings.

Sample cuts

Color or swatch cards sent by textile companies help designers and merchandisers to make fabrication decisions. To test the fabric, the designer orders a three to five-yard cut of a fabric to make a test garment. If the designer and merchandiser are very enthusiastic about a fabric, they may initially order enough for many samples—perhaps one hundred yards or more.

Once fabrics have been selected, the designer can begin to create styles for the new line.

Line

After selecting the fabric, the designer must consider the other design elements. In this section, the term *line* refers to the direction of visual interest in a garment created by construction details such as seams, openings, pleats, gathers, tucks, topstitching, and trims (it is confusing that the apparel industry also uses the term *line* to refer to a collection of garments). Line direction should flow from one part of the garment to another and should not be meaninglessly cut up.

Straight lines suggest crispness, such as that of tailored garments; curved lines imply buoyancy. However, a garment designed with only straight lines is too severe; a garment with all curves is too restless. Curves need the steadying influence of straight lines. For optimal beauty, the two should work together. Straight lines are softened by the curves of the body, and full curves must be restrained to be compatible with the human form.

Lines have the power to create moods and feelings. Vertical lines remind us of upright, majestic figures and suggest stability. Horizontal lines are like lines at rest; they suggest repose, quiet, and calm. Soft, curving lines express grace, and diagonal lines imply powerful movement and vitality.

Shape

Another function of line is to create shape. We use the term *silhouette* to describe the outline of the whole garment. Because the silhouette is what we see from a distance, it is responsible for one of our first impressions of a garment. Silhouettes tend to repeat themselves in cycles throughout history. At times a more body-conscious, natural (hourglass) silhouette is popular. At other times rectangular, inverted triangle or tubular shapes that de-emphasize body contours may be prevalent.

A silhouette should be related to body structure, but some variation is needed to add interest. Sometimes one part of the silhouette, such as sleeves, hips, or shoulders, predominates. Of course, the final effect is heavily influenced by the fabric used. *Bodies* (industry jargon for silhouettes) that do well in one season are usually updated in a new fabric or color for the next season.

Lines also divide the total area into smaller shapes and spaces by means of seams, openings, pleats, and tucks. A good silhouette is composed of parts that in themselves have interesting shapes. Parts may be square, round, oval, rectangular, triangular, or irregular. The minute a waistline is added, a garment is divided into two new shapes: a bodice and a skirt. The sleeve becomes another shape. These parts create new spaces for smaller details, such as collars and pockets. The pattern of the fabric can create even smaller shapes.

Design Principles

Whether design elements are used successfully depends on their relationship to one another within the garment. *Design principles* serve as guidelines for combining elements. Designers may not consciously think of these principles as they work, but when something is wrong with a design, they are able to analyze the problem in terms of proportion, balance, repetition, and emphasis to create a harmonious design. These principles are flexible, always interpreted within the context of current fashion trends.

Proportion

Proportion is simply the pleasing interrelationship of the size of all parts of the garment. When conceiving a style, the designer must consider how the silhouette is to be divided with lines of construction or detail. These lines create new spaces, which must relate in a pleasing way. Generally, unequal proportion is more interesting than equal. Many mathematical formulas have been proposed as guidelines, but the best results come from practice in observing and analyzing good design. Standards of proportion change with fashion cycles along with the evolution in silhouette and line.

The height and width of all parts of a design must be compared. Individual sections of a garment, such as sleeves, pockets, and collar, must all relate in size to each other as well as to the total silhouette. A jacket must have a pleasing visual relationship with the length and shape of the skirt or pants.

Background space is just as important as the detail or shapes within it. A large, bold shape against a plain background is dramatic. Areas broken into small shapes suggest daintiness. Each detail or shape within the silhouette should complement the whole.

An intriguing shape of jacket and neckline and perfect proportions make this Richard Tyler suit special. *(Courtesy of Richard Tyler)*

The spacing of trimmings, pleats, and tucks must have meaning in relation to the total design. Trimmings must not be too heavy or too light, too large or too small to harmonize with the space around them. Every line, detail, or trim changes the proportion because it breaks up the space even more. The designer continually experiments with subtle variations in proportion: line placement, hem length, and size and placement of trims.

Balance

Balance refers to "visual weight" in design. A garment must be balanced to be visually pleasing.

Symmetrical balance

If the design composition is the same on both sides of the garment, then the design is considered *symmetrical* or formally balanced, following the natural bisymmetry of the body. Just as we have two eyes, two arms, and two legs, a symmetrical garment must have exactly the same details in just the same place on both sides. Formal balance is the easiest, most logical way to achieve stability and is therefore, the most commonly used in design. Even slight deviations, when minor details are not exactly alike on both sides, are considered *approximate symmetry*. A sensitive use of fabric, rhythm, and space relationships is needed to keep a symmetrical design from being boring.

Asymmetrical balance

To achieve a more exciting, dramatic effect, asymmetrical or informal balance can be used. Asymmetrical design composition is achieved by a balance of visual impact. A small, unusual, eye-catching shape or concentrated detail on one side can balance a larger, less imposing area on the other side. Striking line, color, or texture can appear to balance larger masses of less significance. Technically, asymmetrical designs make pattern layout more difficult and therefore more expensive.

Repetition

Repetition, or a sense of movement, is necessary to create interest in a design and to carry out the central theme. This can be achieved by the repetition of lines, shapes, and colors to give direction. We can see rhythm of lines and shapes in the repetition of pleats, gathers, and tiers, and in rows of trim,

This Giorgio Armani design uses resourceful repetition in the folds of the skirt, fringe trim, and combination of striped fabrics.
(Courtesy of Giorgio Armani)

Another Giorgio Armani design features an ingenious draped and twisted focal point. *(Courtesy of Giorgio Armani)*

banding, or buttons. The dominant color, line, shape, or detail of the garment may be repeated elsewhere with variation. The sense of movement must be felt, even if subtly.

The use of repetition is one of the most helpful guidelines in designing. A design line, shape, or detail repeated in another area of the garment helps to carry the theme throughout the whole design. In dress design, for example, a V neck might be repeated upside down in bodice seaming or in an inverted pleat in the skirt. Soft gathers at the neck could be repeated at the hip to unify the design.

Emphasis

Emphasis, or a center of interest, draws attention to the focal point of a garment. This point is the central theme; the rest of the garment is of secondary importance. A center of interest must create more visual attraction than any other design element in the garment, and all other elements must support it by echoing its design message with weakened impact. A center of interest should be related to the total structure of the garment. Such a focal point can be achieved by accenting with color, significant shapes or details, lines coming together, groups of detail, or contrast. A combination of these methods gives the focal point added strength, as does placing the decorative emphasis at a structural point. Karl Lagerfeld does it simply with rows of gold buttons for Chanel.

In the designing of sportswear, each piece is considered a part of the whole. Although some pieces are simple basics, they create a background for or complement another piece such as a jacket which creates the center of interest.

Above all, the wearer should be the center of attraction. Lines leading to the face are particularly effective. Light colors around the face also help direct interest to the wearer. This is one reason why contrasting collars are common.

Successful design

A successful design is achieved when all the elements and principles of design work together harmoniously. The theme of the garment is carried out with nothing overdone or forgotten.

An effective design results from a well-developed idea or theme. For example, if the theme of a design or group is dramatic, the design should have a bold statement of line, an exaggerated silhouette, large space divisions, bright or dark colors, strong contrast, large prints, or extreme textures. There are many ways to develop ideas and themes.

A well thought out use of the elements and principles of design is most apparent in an evening gown when the drama of the occasion makes it appropriate to create something sensational. For daily wear, however, garments are simpler and elements less noticeable.

The best way to gain experience in design is by experimentation. A designer on the job is always learning. The designer often tries many variations of a design before creating one that has the perfect combination of fabric, color, line, and silhouette and the correct use of balance, proportion, emphasis, and repetition.

Designers usually work up their ideas in sketch form to test them. The designer must determine objectively whether all the elements work together to create a harmonious, consistent visual effect.

Sketching Ideas

Ideas sometimes originate on the drawing board. Lacroix admits, "I pile up hundreds of drawings."[8] Starting with an idea for a silhouette or a neckline, the designer may experiment, sketching alternate ways to complete the design. On paper, the designer can see two-dimensionally which design elements might enhance one another. (The designer must also be able to imagine how the garment will look three-dimensionally, when made up in a fabric.)

The first rough sketch, a working sketch, is very simple. However, proportions must be accurate so that the patternmaker can interpret it correctly. The designer must think out the garment's construction in relation to the sketch, which must be drawn in proportion to the body,

Designers can experiment with color and pattern changes by computer.
(Courtesy of Shima Seiki USA)

showing exact details of seam line and trimmings. Sketches must be clear so that the patternmaker can use the drawing as a guide. The working sketch usually includes notes on construction and measurements. Fax machines have been a big help in letting designers communicate their design ideas quickly to overseas offices and production facilities.

Computer-Aided Design (CAD)

Designers can also sketch on a computer. So far, computer-aided design is used primarily to show how a garment might look made up in fabric. Computer graphics show the repeat of a pattern on the sketched garment. The system can also simulate fabric drape on the figure and help designers experiment with color and fabric changes quickly. Computer companies are also experimenting with the idea of converting sketches automatically into first patterns.

Style Board

To chart the development of the line as a whole, the designer arranges working sketches of all the garments in fabric and color groups on a large board, which is essentially a master plan (see the photo of Christian Lacroix at the beginning of the chapter). The board is posted on the wall of the design room. Styles are added and subtracted until the line is established. The line may also be organized and edited by computer.

DEVELOPING A SAMPLE GARMENT

The First Pattern

The next step in the product development procedure is making the first pattern, which is used to cut and sew the sample garment. The pattern is made in a sample size, the one used for testing and selling purposes. Sample sizes are 7 for juniors, 8 or 10 for missy, 34 for men's trousers, and 38 for men's suits. The patternmaker can use either of two methods for making patterns: draping or flat pattern.

Draping

The patternmaker or assistant designer uses the *draping method* to cut and shape muslin or garment fabric on a dress form to create a pattern. Like fabric sculpture, draping is ideal for soft, flowing designs. This enables the patternmaker and designer to see the proportions and lines of the design exactly as they will look three-dimensionally, on a body. The design is often altered as it develops on the form.

Everything must be carefully marked: center front, shoulder line, seams, armholes, buttonholes, and so on. When the designer is satisfied with the look and fit, the muslin is taken off the form and trued up. *Truing* is correcting the sketched lines with drafting curves, angles, and rulers. Finally, the pattern is redrawn on heavy paper.

Flat pattern making at Jessica McClintock's design studio. (*Photographed by the author*)

Draping a pattern at Jessica McClintock's design studio. (*Photographed by the author*)

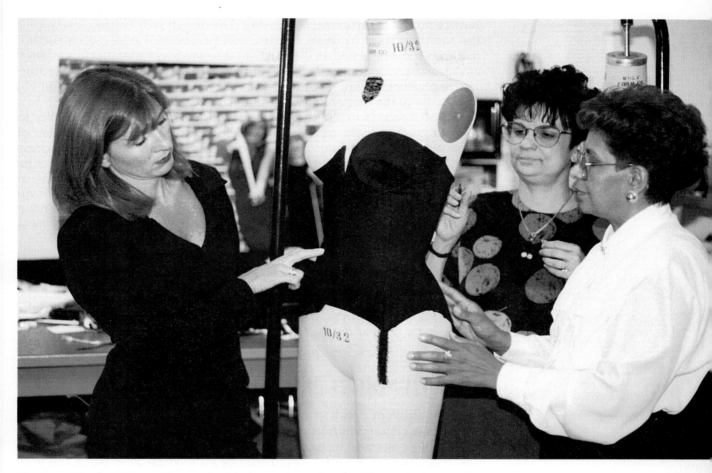

Designer Nicole Miller, with samplemakers, works on the development of a sample garment. (*Courtesy of Nicole Miller, photographed by the author*)

Flat pattern

The *flat-pattern method* uses angles, rulers, and curves to change existing board patterns. First, basic shapes, such as bodices, sleeves, pants, or skirts, are draped or drafted. *Drafting* is blocking out a set of prescribed measurements on pattern paper for each piece. These patterns, once tested for accuracy, become *blocks* or *slopers* that can be changed or adapted to each new style by moving darts and seams. This method is necessary for areas like sleeves and pant legs, which would be difficult to drape on a figure.

Many patternmakers combine the two methods, depending on the dictates of the design and on their technical training, skill, and preference.

Sewing and Fitting the Sample

After the pattern is made from Manila paper or tag board, it is laid out on the fabric, traced, and cut out by the assistant designer or the sample cutter to be sewn into the first sample or prototype.

The prototype is made by a sample maker, the best of the factory sewers. Sample makers must have factory sewing skills but also know how to put an entire garment together. The design staff works closely with the sample maker to solve construction problems. Factory construction methods must be tested as the garment is sewn. Some manufacturers, such as Adrienne Vittadini, have their samples made abroad at the same factory where production will be done due to the need for specialized equipment (especially in the case of knitwear) and to test factory production.

The next step is to test the sample for fit and total effect. The ability to create a good fit is the most important skill needed in the development of the garment. It is necessary to see how the garment fits not only on a dress form, but also on a model to test comfort and ease of movement. Donna Karan likes to try on her samples herself. The designer must be objective in judging the design and develop an unbiased, critical eye—a difficult task when judging a personal creation. Adrienne Vittadini feels that the hardest lesson she had to learn was "detaching myself from the emotional tie—that each garment is your baby."[9]

Duplicates

Duplicates are copies of the sample garments that are made to keep in the showroom to show to buyers or to give to sales representatives to take on the road. Duplicates are also useful as guides in factory production.

The Designer Work Sheet

Records are kept on all styles as they develop. Each designer fills out a work sheet containing information that guides the production department in figuring costs and in ordering *piece goods* (the factory term for fabrics) and trimmings. The work sheet becomes an *adoption sheet* when the final line is selected.

The work sheet includes the following information:
1. The date the garment was designed.
2. The selling season for which the garment is designated.
3. The sizes in which the design will be made.
4. The style number assigned to the design. Each manufacturer has a code of numbers representing season, fabric, and pattern.
5. A short description of the garment.
6. A working sketch of the design, to make it easy to identify the garment.
7. The colors or color combinations in which the design is offered.
8. Fabric swatches of what is used in the garment.
9. Material descriptions, including each fabric type and source, width, and price per yard.
10. Marker width, usually one inch narrower than the fabric
11. Trimmings information: the kinds, sources, sizes, and prices of buttons, zippers, braids, lace, belts, and elastic. Special fabric treatments done by outside contractors, such as pleating, spaghetti straps, and ties, are also included.
12. Labor costs for grading, marking, cutting, and sewing. These may be listed either on the designer work sheet or on a separate cost sheet figured by the costing department. See Chapter 10.

Designer Work Sheet

DATE				STYLE No. *1025*
DESCRIPTION *Floral print dress*				SEASON *spring*
				SELLING PRICE: $ *61.52*

SIZE RANGE *4-14*   COLORS _____

MARKERS _____

MARKER YARDAGE: _____   ALLOWANCE: _____

SKETCH

1. MATERIAL

MATERIAL	YARDS	PRICE	AMT.
print	4.75	3.85	18.29
Lining			
TOTAL MATERIAL COST			*18.29*

2. TRIMMINGS

TRIMMINGS	QUANT.	PRICE	AMT.
Buttons *#20 covered*	6	.03	.18
Pads			
Embroidery			
1½" wide satin ribbon	1.7	.25	.43
Belts			
Zippers			
Pleating, Tucking			
¼" elastic	.7	.05	.04
interfacing	.15	1.05	.16
fusing			.50
label			.12
shipping			.08
TOTAL TRIMMINGS COST			*1.51*

3. LABOR

LABOR			
Cutting			2.00
Labor			
grading and marking			.75
sewing $126.00 per dozen			10.50
pressing			.75
Payroll Taxes & Health Fund			.46
Trucking			
TOTAL LABOR COST			*14.46*

4. TOTAL COST $ *34.26*

REMARKS

LINE SELECTION

Regular *line-in-process* meetings are held to analyze the development of the line. Designers develop many designs for each line which are then edited. From all of the samples, management, the designer, and the merchandiser choose the best for the line. Successful buyers and sales representatives are often asked for their opinions because of their familiarity with customer preferences and retail merchandising expertise.

Some styles are weeded out, leaving only the most successful combinations of fabric, style, and price. Some companies feel that they need a closely edited line so that combination possibilities are not confusing to the customer. Other large companies like to give their customers more options. Escada, for instance, typically shows 600 pieces per collection; Adrienne Vittadini, 200 pieces; and Ellen Tracy, approximately 160 pieces. Since the recession, collections have tended to be carefully edited to reduce the number of garments.

At this point, merchandise plans might be *re-assorted*. If a group is particularly strong, styles or colors could be added. The dollar merchandise plan we discussed at the beginning of this chapter could be revised so that it states the exact number of styles adopted, the actual price of fabrics chosen, and the yardage estimates for each style.

Just before the scheduled line-release or collection-opening dates, the line is presented to the sales force. The designer or merchandiser must explain the line concepts to the sales representatives, who can use this background information when showing the line to retail buyers.

A consistent visual image created by a well-developed theme is a good selling tool on both the wholesale and the retail levels. Visual impact is what makes the line competitive once it is displayed in a retail store. The designer cannot be there to explain each style idea to the customer, so the garment design must speak for itself. Design development is a challenge. Not only must designers be creative; they must understand production and also know what sells. The ultimate blame for a bad season (assuming that economic conditions and the sales force remain the same) falls on the designer.

SUMMARY

In this chapter we discussed product development, merchandising and design. The company establishes an image and a product to appeal to a target market. The merchandiser prepares plans to achieve sales goals and meet consumer needs. The design department creates a seasonal line of items or groups. The designer considers color, fabric, line, shape, proportion, balance, rhythm, and focal point in each design in order to develop pleasing garments. After a sketch and first pattern are complete, the sample garment is made as a test of the design and fabric. Finally, the line is selected and prepared to show in the wholesale market.

CHAPTER REVIEW

Terms and Concepts

Briefly identify and discuss the following terms and concepts:

1. Collection or line
2. Seasons
3. Merchandising
4. Dollar merchandising plan
5. Designer
6. Items
7. Knockoffs
8. Planning groups
9. Theme
10. Elements of design
11. Color story
12. Fabrication
13. Texture
14. Performance
15. Sample cuts
16. Proportion
17. Asymmetrical balance
18. Repetition
19. Working sketch
20. First pattern
21. Draping
22. Flat pattern
23. Sample garment
24. Designer work sheet
25. Style board
26. CAD

Questions for Review

1. Name the three major divisions of a clothing company, and discuss their relationship.
2. Discuss the merchandising function in relation to the design function.
3. Discuss the basic conceptual differences between merchandising coordinated sportswear and merchandising separates.
4. Name the elements of design. Discuss briefly why it is important for these elements to be represented in a design.
5. What are the principles of design? How do they help a designer analyze the effectiveness of a design?
6. Explain briefly the development of a sample garment.

Projects for Additional Learning

1. Write a short description of an imaginary or real person. Include the person's age, build, job, place of residence, interests, and life style. This person will be the typical customer for your new sportswear line for the season eight months from now. Determine a price range and style range according to your customer's lifestyle. Design a group of twenty coordinating pieces or related separates. Develop a color story and suggest fabrications. (Merchandising students may cut pictures of sportswear items from magazines and display them on a board.)

2. Find examples of ten successful designs in European and American fashion magazines. Describe the beauty of each in terms of line, silhouette, shapes, spaces, color, and texture (fabric, pattern, and trims) as well as balance, proportion, rhythm, and emphasis . Do all elements and principles work together harmoniously? What elements dominate? Find two garments that you feel are unsuccessful. Analyze them in the same way to discover what element or principle has not been properly used.

NOTES

[1] "Lacroix Tells Us About Christian," special supplement to *Marie Claire* (October, 1988), p. 1.

[2] As quoted in "Lauren at 25," *Women's Wear Daily,* January 15, 1992, p. 7.

[3] Ibid, p. 6.

[4] Interview with Robert Abajian, Design Director, Liz Claiborne, May 1, 1992.

[5] As quoted by Nicholas Coleridge, *The Fashion Conspiracy* (London: Heinemann, 1988), p. 144.

[6] Abajian, May 1, 1992.

[7] Christian Dior, *Talking about Fashion* (New York: Putnam's, 1954), p. 35.

[8] "Lacroix Tells Us About Christian," p. 1.

[9] Quoted by Lesley Nonkin, "Vittadini al dente," *Women's Wear Daily*, April 9, 1980, p. 6.

A cutter at work at Jessica McClintock's factory. *(Photographed by the author)*

10

APPAREL PRODUCTION

CAREER FOCUS

There is an interesting variety of job possibilities in apparel production. There are 20,000 apparel production plants throughout the United States. Entry-level positions are usually available in shipping and quality control. Technical training in both patterning and computers is needed for positions as patternmakers. Engineers are needed for computer-aided manufacturing. For those interested in materials, there are positions as fabric and trimmings buyers. Numbers-oriented people might like to work in cost analysis and production planning. Experience brings promotions to managerial positions in each production area.

189

CHAPTER OBJECTIVES

After reading this chapter you should have attained competence in the following areas:
1. The ability to explain the costing of a garment.
2. The ability to describe, in order, the steps in garment production.
3. Understanding of the various types of contracting.
4. Understanding of all of the uses of computers in manufacturing.
5. The ability to explain the differences between men's tailoring and men's sportswear production.
6. The ability to describe the two commercial methods of knitwear production.
7. Understanding of the importance of quality control.

*P*roduction is another of the three integral phases of fashion manufacturing: design, production, and sales. Without design, there would be nothing to sell; without sales, there would be no reason to produce. Conversely, sales would not continue if orders were not filled properly and delivered on time. This chapter focuses on apparel production, from patternmaking to quality control.

After the collection or line is shown at a market, orders are sent to the factory. Styles that receive insufficient orders are dropped from the line. To keep its retail customers, each manufacturer tries to maintain a consistent price structure, quality of styling and construction, and timely deliveries.

COSTING A GARMENT

The production cost of a garment must be determined in order to set the wholesale price, the price that retailers pay for goods they purchase from manufacturers.

Costing Functions

There are two separate costing functions: the precost and the final costing.

Precost

The *precost* is an estimate made before the garment is adopted into the line. From the outset, the designer must keep fabric and labor costs for each garment within the limits set by the company for a particular line's price range. The designer keeps a record of all material costs on a designer work sheet. Then, either the designer or the costing department can roughly estimate the wholesale cost to determine whether the garment fits into the line's price structure.

Final costing

This is an exact calculation by the costing department, utilizing final figures for materials and labor. The costing department uses the designer's work sheet, an actual sample, and the production pattern to analyze the garment's materials and construction step by step. It may consult the designer for information or recommend more practical or cheaper ways to make the garment. A detailed cost analysis is made for each garment, including expenses for fabric, trims, cutting, labor, overhead, sales commission, and manufacturer's profit. The final cost is plotted on a cost sheet. It is now much easier to do this on a computer.

The Cost Sheet

1. **Materials.** First, the total amount of yardage of each fabric needed for the garment is estimated and then multiplied by its cost per yard. The sum of these figures is the total material cost per garment. Higher volume allows more flexibility in making the marker and thus a more efficient use of fabric. Material cost is somewhat reduced for high-volume styles while extra expense has to be allowed for a low-volume style.

2. **Trimmings.** Unit costs are multiplied by the amount of trimmings needed for each garment. The sum of these figures is the total trimmings cost per garment.

3. **Production patternmaking, grading, and marking.** Most companies allow for these costs in the general overhead that also covers the design department. However, if these functions are performed outside by a pattern service, the cost is divided by the total number of units they estimate will be cut plus a profit for the contractor:

   total cost ÷ units to be made = cost per unit

   If the garment is later recut (because of reorders), there will be no new cost for patterns, grading, and marking.

4. **Cutting.** The cost for cutting done in house is based on the cutter's hourly wage multiplied by the number of hours it takes to cut the style.

   If the cutting is done by a contractor, the total negotiated cutting cost is figured on the number of garments to be cut. The contractor adds his or her fee to this amount.

5. **Assembly.** Construction labor includes all sewing including finishing. Some companies break down labor costs by each operation. Information for such a costing structure is gathered through time-and-motion efficiency studies. The cost for each operation, such as the closing of a shoulder seam, is determined. To figure the total costs for a whole garment, the individual operation costs are added.

   Other firms calculate the average time it takes to sew the whole garment and multiply that by the worker's hourly rate.

TABLE 10-1
Manufacturer's Cost

Wholesale Unit Pricing for a Typical Dress
(each company uses a unique pricing structure)

Direct Cost	Amount	Average Percentages
Fabric (cost depends on quanity purchased)		
4.75 yards at $2.48	$18.29	30%
Trimmings	1.51	2
Labor (cost depends on number of garments sewn)	14.46	24
Total Direct Cost	34.26	56%
Indirect Cost		
Design and Merchandising:		
Design Staff salaries, Sample fabrics, Cost of samples	5.00	8 (8-10)
General Administrative Overhead:		
Office Salaries, rent, insurance, utilities	5.28	9 (8-15)
Sales Commission	4.50	7 (7-10)
Trade Discount	5.00	8
Markdown allowance, promotion, or other retail services	2.40	4
Shortages	0.96	2
Total Indirect Cost	23.14	38
Total Cost	57.40	93
Profit before taxes	4.12	7
Wholesale Price of Dress	$61.52	100%

(refer back to design chapter for work sheet on same dress; see retailing chapter for retail price of the same dress)

If the garment is made by a contractor, the contractor's fees must be added to production costs.

6. **Finishing.** Costs must be included to cover final preparations. In designer apparel, this may include some hand work. For most apparel, washing, pressing, and folding is done in house. The cost for packing in preparation for shipping is also figured.

7. **Freight.** The cost of shipping completed garments from the contractor to the manufacturer must be calculated. For domestic shipments, the garments are usually trucked. If the garments are imported, then a percentage of the air or sea freight cost must be added to the cost of each garment. Obviously sea transportation is cheaper and therefore adds less cost to the garment, but valuable lead time is forfeited.

The cost of shipping garments to the retailer is generally paid by the retail store (the receiver). But manufacturers must pay air freight if they are late with their delivery.

8. **Additional costs for imported garments.** These include the quota charge and/or import duty, and agent fees.

Wholesale Pricing

The wholesale price is determined by adding the cost of labor, materials, and a markup. The markup covers *sales commission,* (usually 7 to 10 percent), *terms* (an 8 percent discount to the retail stores if they pay their bills on time), overhead, and a profit, which is necessary to stay in business. *Overhead* includes all the daily costs of running the business. *Fixed* overhead expenses include rent, heat, lights, trash collection, insurance, salaries, taxes, and legal and auditing services. *Variable* overhead expenses include advertising, promotion, equipment repair, markdowns of leftover fabrics and garments, and losses due to fire and theft. Companies also have to use a large portion of their profits for capital investment in new equipment and technology.

Most companies add a markup of between 28 and 44 percent of the wholesale price. In other words, they increase the direct cost by 56 to 88 percent. Budget companies usually take the lowest markup and hope to make a profit by volume sales. For example, a blouse costing the manufacturer $21 to produce, with a 40 percent markup would cost $35 wholesale.

Cost Merchandising

Another important aspect of costing is whether the garment's perceived value is worth its price. Each garment has to compete on the market with other, similar garments. If two garments are very much alike, the less expensive one is more likely to sell on the retail level. Therefore, both the styling and quality of a garment must be better than those of competing garments.

The price of a garment may be adjusted slightly higher or lower to try to affect its sales potential. Occasionally a manufacturer uses a *loss leader* (a garment with a low markup) to attract buyers.

Some companies figure an *average markup* so that their garments fit into a price line. For example, the consumer is usually willing to pay more for skirts and pants than for shirts. Yet shirts often cost out higher because of expensive fabrications and labor. Therefore, the shirt price may be lowered so that it is more acceptable (a practice called *low-balling*), while the skirt and pant prices are raised *(high-balling).* In this way the average markup is within an acceptable range.

Many volume manufacturers *re-engineer* some of their first samples to meet their target price points, if the garment in question is important to the line. This is accomplished by reducing fabric consumption by altering such features as the sweep of a hem or facings or pockets.

PURCHASING OF PIECE GOODS

Ordering the materials necessary to produce garments is usually done by the manufacturer's fabric or piece goods buyer.

The *piece goods buyer* acts as a liaison between the mill or converter and the manufacturer, who must have a good credit rating in order to purchase the fabrics. The buyer must know all the properties of fabrics and have information on prices, availability, and delivery. Some things that must be considered in the purchase of piece goods are as follows:

Environmental concerns. Consumer preference for environmentally safe products has led some manufacturers to look for naturally colored cotton; organically grown cotton, linen, wool, and tencel; vegetable-based and water-based inks for prints; fiber-reactive, low-impact dyes; and safe finishes for the fabrics they purchase.

Volume purchases. Generally, volume manufacturers purchase goods from large textile companies that can handle big orders. Ordering in quantity may give the manufacturer a cost advantage. The goal of a volume manufacturer is to obtain the lowest-priced goods at the widest width, in order to use an efficient marker layout.

Smaller orders. Conversely, small manufacturers usually deal with smaller textile firms, which can do shorter runs. This often forces American companies to buy imports because their minimums are lower. For example, a manufacturer may be able to buy a minimum of 300 yards of a fabric from Asia versus a 2000-yard minimum in the United States. Many manufacturers also argue that quality and dye lot consistency is better in Europe or Asia as well.

Ordering. The buyer must calculate the amounts of yardage needed to supply the cutting order. Orders for stock yardage must be based on sales histories and on expectations of future sales. Textile sales representatives encourage manufacturers to make a commitment early in the season to buy a specific amount of yardage. Ordering in advance ensures the availability and on-time delivery of a popular fabric, which may subsequently be in short supply. Volume orders usually require a seven-week to eight-week lead time for delivery. Popular fashion fabrics and yarn dyes can require as much as a six-month lead time.

Fabric for reorders. The manufacturer must also consider the amount of fabric that may be required for possible reorders. A manufacturer takes a great risk by stocking up for anticipated orders. To eliminate some

Evolution of a Garment

Fiber
↓
Mills
↓
Weaving or knitting | Dyeing | Finishing (Special Processes)
↓
Manufacturer's Purchase of Sample Cuts
↓
Designer → Pattern Maker → Sample Maker
↓
Salesman's Samples | Showroom
↓
Showings to Buyers
↓
Fabric Purchase | Pattern Grading
↓
Cutting → Sewing → Finishing → Pressing
↓
Quality Control
↓
Shipping
↓
Retailer
↓
Consumer

risk, manufacturers often commit for *greige goods* only, to be dyed later. Then, as sales information becomes more precise, colors can be *assorted*. In other words, the greige goods can be dyed in various colors according to the sales percentage established for each. Designer and bridge fashion manufacturers are usually unable to reorder materials because the fabrics are imports. There is not enough lead time to wait for another fabric shipment.

Trimmings. The trim buyer buys trimmings and findings. These are the materials used to finish garments or fashion accessories. *Findings* are functional materials such as linings, zippers, thread, and seam tape. *Trimmings* are decorative and include such items as buttons, braids, laces, and belts. Trimmings must have the same care properties as the fabrics used in the garment. For example, if the fabric can be safely washed, so must the trimmings. As trimmings and findings arrive at the plant, they must be sorted and inspected along with fabrics. Other sundries purchased by the trimmings buyer include labels, hang tags, hangers, and plastic bags.

Piece Goods Inspection

As fabric is received in the garment factory, it must be carefully measured to make sure the total order has been received. Special equipment automatically counts the fabric yardage as it rolls from one bolt to another. The fabric is then pulled by rollers over large viewing tables or examined on the cutting table to check for flaws such as holes or shading. Flaws are marked with colored threads or flags at the side of the fabric so that they can be avoided in cutting. In an automated factory, a computer-programmed inspection system locates flaws, color differences, and variations automatically. If there are too many defects, the fabric is returned to the mill.

Some apparel manufacturers build strong reputations on the reliability of their products. They perform washing, dry-cleaning, steaming, and pressing tests to check tensile strength, durability, color cracking, color fastness, and shrinkage. They also test how the fabric holds up during sewing—whether it pulls or frays excessively.

Coordinating tops and bottoms must be cut from the same dye lot and, at the very least, shipments to the same store must be within acceptable variations of color if from different dye lots.

COMPUTERS IN MANUFACTURING

In automated factories, computer technology is used in every phase of apparel manufacturing. Computers control inventory and make production faster and more efficient, but require enormous financial expenditure.

Inventory Control

The introduction of computerized inventory control has made planning more accurate. Computers are used to create issue plans and cutting orders. All fabrics and work-in-progress are given *universal product codes*

(UPC), which identify style, color, size, price, and fabrication. Computers provide current information on the number of units that are cut, in progress, or in stock. With these data, manufacturers can update fabric commitments and production plans so that they correspond with demand.

Quick Response

Quick Response strategies were started by the domestic computer and textile industries in order to quickly replenish stock, giving the manufacturer a competitive edge against imports. Quick Response requires the use of the newest computer technology to eliminate wasted time in the textile-apparel-retail pipeline.

Electronic Data Interchange (EDI)

Quick Response is carried out by cooperative alliances between all levels of the industry via *electronic data interchange,* the exchange of business data between two or more parties by computer. Linkage systems from one level of the industry to another and standardized codes for data exchange throughout the industry have been established. These bar codes are laser scanned and the information is fed through the EDI pipeline.

When a particular style is selling well at retail, coded information on that style, including the best-selling colors, fabrics, and sizes, immediately informs the retail buyer to replenish stock, the apparel manufacturer to issue new cuts, and the fabric producer to send more fabric to the apparel manufacturer.

Of course, for this strategy to work, it needs the complete cooperation of each level of the industry: a willingness to supply on a reorder basis, the flexibility to sell smaller quantities of fabric, and issue smaller, more frequent cuts. Issuing smaller initial cuts until a style is tested prevents markdowns and unnecessary inventory while quickly replenishing stock with best-sellers will mean higher sales. This allows product development to be done closer to the selling season in order to better anticipate consumer preferences. Quick Response works well for basic apparel and now even some fashion manufacturers such as Nicole Miller are using it too.

Patternmaking and Garment Assembly

Computers are also used in every phase of patternmaking, cutting, and garment assembly. This information is discussed in the appropriate sections.

PATTERNMAKING

Accurate patternmaking is crucial for successful apparel production.

The Production Pattern

In a traditional apparel factory, the production patternmaker relies on the same methods used in the sample room to make patterns: draping, drafting, or flat-patterning from standardized basic blocks. In fact, in many small companies the same person does both the sample and production

A pattern maker using Gerber's AccuMark Silhouette pattern development system. The pattern maker works on a backlit pattern making table with a stylus that automatically records lines in the computer and shows them on the screen. *(Courtesy of Gerber Garment Technology, Inc.)*

patternmaking. In production, strict attention must be paid to company size specifications, which are standardized measurements including *ease* (extra room for movement) for each size. When using fabrics that shrink, such as cotton, patterns have to allow for that shrinkage. Also, the edges of each pattern piece must exactly match the piece to which it will be sewn, with notches marked perfectly for operators to follow.

Computer pattern making

In an automated factory, patterns are made on a computer. With *computer-aided-design (CAD)* systems, the pattern maker manipulates small graphic patterns on the computer screen with a hand-held control device. Geometry drivers can make 100 changes to the shapes and sizes of the patterns, including creating new design lines, or adding pleats, fullness, and seam allowances.

To allow patternmakers to work with their hands, another system has been developed allowing the patternmaker to work lifesize on a sensitized table with traditional tools and a stylus that is attached to the table and the computer. The stylus picks up the lines drawn on the table and shows them on the screen. Changes can also be made directly on the screen.

In both cases, patterns are immediately available for other operations such as grading and marker making.

Grading Sizes

Patterns, like garments, must provide for different sizes. *Grading* is the method used to increase or decrease the sample-size production pattern to make up a complete size range. For example, the sample size 10 pattern must be made larger to accommodate sizes 12, 14, and 16, and made smaller for sizes 8 and 6. Each company sets predetermined grade specifications, or rules. For example, a missy manufacturer's grade rules might call for increments of one and a half inches in width and a quarter inch in length for each size.

Traditionally, the sample-size pattern is held in place over tag or manila board by a grading machine. The operator turns one knob of the machine to move the pattern forward and backward and the other knob to move the pattern sideways to mark points of size change. The pattern is moved for each point of increase or decrease. The patterns are hand-traced by the operator. Each size may be made on a different color pattern board for easy recognition.

Computerized grading

In the automated factory, the patterns are graded by a CAD system on a computer. The operator guides a digitizer around the edges of the sample pattern. At each of the key points, he or she pushes a button to record a grade point. Each point is cross-referenced by the computer digitizer to a grade rule, which enlarges or reduces the pattern automatically according to predetermined increments and in a predetermined direction.

If the pattern was originally made by computer, the data is already stored and can be enlarged or reduced automatically. The computer then prints out the pattern in each new size.

Making the Marker

From all the pattern pieces of varying sizes, a master marker is made. The *marker* is the cutting guide or pattern layout, made on a sheet of lightweight paper the same width as the fabric. The purpose of the marker is twofold: to place pattern pieces close together to avoid fabric waste and to accommodate the cutting order. The desired economical use of space is called a *tight marker* which utilizes the highest percentage of fabric possible. To accommodate the cutting order, patterns are laid out so that each size and color are cut as needed.

Traditionally, manila board patterns, already graded, are traced onto marker paper. Grain direction, one-way prints, and naps are considered in making the marker. Copies of the marker are made to use in each cut.

Computer markers

In an automated plant, a CAD system makes the marker. Miniatures of the graded pattern pieces are displayed graphically on the computer screen. The operator can electronically position the pattern pieces into the most efficient arrangement. Once the arrangement is completed, the computer causes the plotter to print out a full-scale marker on a long sheet of paper. Miniatures of an entire marker can be faxed to locations around the world.

PRODUCTION SCHEDULING

The production manager schedules cutting and garment assembly in time to meet shipping dates and organizes operations to take maximum advantage of plant capacity.

Issue Plan

A production schedule or *issue plan* is created on a computer to ensure that delivery dates are met. This schedule is a reverse timetable, usually covering six months. The first date on the schedule is a shipping date that will meet the retail store's order requirements. The schedule progresses backward to include completion dates, cutting dates, and even fabric delivery dates.

Orders for each style must be compiled to determine how many garments of each size to produce. There are two philosophies of production planning: cut to order and cut to stock.

Cut to Order

The safest method of production is to *cut to order*, that is, to cut and produce only against orders. This means waiting until all orders are in and then working quickly to cut, sew, and deliver. As mentioned earlier, an ideal situation for retailers would be to test and reorder. However, this method can work only under model conditions using the newest production technology and with close communications and cooperation between levels of the industry.

Cut to Stock

The method with the greater risk involves cutting against estimates of projected sales or *cutting to stock*. *Projections* (expected sales) are determined by the economy, sales histories of similar garments, the season, and the strength of the particular line. If the manufacturer is confident of a style, or if preliminary orders indicate a big seller, the manufacturer cuts and produces that style to have stock available for fast shipment.

The cut-to-stock method is necessary especially in large companies that produce moderately and popularly priced fashion in volume. It is also necessary when orders for fabrics require long lead times, as is the case with yarn-dyed fabrics. The method enables the company to start work in slack months on those items with the highest projected sales. This provides continuous work for permanent employees and continuous use of plant facilities. During peak production months, the manufacturer will seek additional facilities provided by outside contractors.

Plant Capacity

Plant capacity and construction difficulties must be considered in planning the schedule. For example, some styles are more difficult to produce than others and therefore take more time. Operators are slower at their work when it is new to them; plant capacity builds as operators become accustomed to the sewing operation and therefore work faster. The production process is similar to cooking; one must add in various ingredients and perform different actions in order to have the final results ready by a specified time.

CUTTING PROCEDURES

Using the markers made from graded patterns and in accordance with the issue plan, fabrics are cut to prepare for garment assembly.

Following the issue plan, the *cutting order* tells what to cut, what fabrics to use, and how to cut it. A computerized cutting order utilizes programmed data to determine what markers should be made, how fabrics should be spread and cut, and the most cost-effective number of plies, colors, and size mixes.

Fabric is spread on the cutting table with one ply on top of another so that many layers can be cut at the same time. Sometimes fabric is inspected as it is spread. Cutters must be aware of how fabric finish, pile, stripes, floral designs, and other factors affect their work.

Conventionally, a spreader machine is guided up and down a table, folding back one layer of fabric over another. The length of each layer is determined by the marker and the number of layers by the cutting order. More than 300 layers may be cut at one time, depending on the fabric thickness.

Spreading can be done automatically with a computer-programmed spreading machine. In this case, a motorized spreader moves automatically up and down the table, unwinding up to 100 yards of fabric per minute without distortion.

A computer-controlled spreader which provides distortion-free fabric spreading at speeds up to 80 yards per minute. *(Courtesy of Gerber Garment Technology, Inc.)*

Cutting Techniques

The marker is put on top of many layers of fabric. Traditionally, the cutter follows the pattern outlined on the marker, using a straight-knife machine with a long, thin blade that vibrates vertically as it is pushed through many layers of fabric. A vertical knife can cut to a depth of ten inches or more. For only a few layers, a cutting machine with a rotating circular knife may be used. The cutter must select the correct speed and blade for each type of fabric. For example, a coarse blade edge is used for tightly woven fabrics, a smooth edge for softer fabrics.

Computer cutting

A numerically controlled cutter reduces labor and improves the accuracy of the cut. A beam structure across the table holds the cutting head. The beam can move up and down the table while the cutting head moves across. Movement is directed by data in the CAD system. The knife in the cutting head vibrates vertically to cut the fabric.

Laser-beam cutting is sometimes used for men's suits a single layer at a time. The laser, a concentrated light beam, is also directed by a computer. *Water-jet cutting* is being used for some fabrics and leathers, especially in the shoe industry. A thin stream of water, also computer-directed, is fired under high pressure through a tiny nozzle to cut the leather.

Die cutting

Die cutting may be used for garments or parts of garments that do not change from season to season. A die, a device that operates much like a cookie cutter, is made for each piece to be cut. The sharp edges of the die are pressed against the layers of fabric to cut it. A *gang die* can be made by connecting several dies together.

Automated computer cutting showing the cutting head, beam structure, and multiple-layer cut pieces. *(Courtesy of Gerber Garment Technology, Inc.)*

Cutting by hand

Modern technology notwithstanding, it is quite a contrast to find cutting done by hand in a couture house or in the sample room. Quantity is not necessary in this case, and single-layer cutting makes absolute accuracy possible.

Bundling

The process of sorting cut pieces and tying them together is referred to as *bundling.* Parts of garments and necessary findings must be grouped for the sewing machine operators. Identification tickets are attached for *piecework* control if factory sewing operators are paid according to how many pieces they sew. The style number, bundle quantity, individual piecework rate, and operation to be performed are printed on these tickets. Bundled work is distributed to machine operators in the factory or sent out to contractors to be sewn.

CONTRACTING

Manufacturers are able to purchase patternmaking, cutting, and sewing from outside services or factories.

Most manufacturers do not handle the entire production of a garment in their own factories. They are responsible for all phases of manufacturing, from design and fabric purchase to selling and shipping, but they may contract out some or all of the production. Just as a manufacturer can purchase designs from a design service, it can also contract out patternmaking, cutting, and sewing.

Many manufacturers contract out all sewing because they do not own any production facilities. These manufacturers do not have to pay wages during slack seasons or to be concerned with hiring, training, or wage demands of personnel. They also do not have to invest in plant facilities and machinery which require large capital investment.

A *contractor* is an independent producer who does sewing for manufacturers. Contractors are hired only as production is needed and they have equipment and trained operators for specialized work such as sewing knits, pleating, quilting, embroidery, or piping.

Contracting provides greater production flexibility, but it can involve problems. The manufacturer has less control over quality. There is extra movement of goods which could result in extra costs and some losses. There may also be communication problems or a possiblity of late deliveries.

The production manager selects the contractor who is most reliable and best suited for a particular job. It is very important that the production head work closely with the contractor to make sure that standards and time schedules are met. Most contractors specialize either in a particular quality of sewing or in sewing one type of fabric such as knits or wovens. Some contractors may arrange to work exclusively with one or more manufacturers. A contractor agrees to maintain a certain standard of workmanship and to finish work by a specific date. The contractor is given cut work, a sample duplicate, and specifications to follow. Finished garments are returned to the manufacturer for shipping to retail stores. Manufacturers may use domestic or foreign contractors.

Global Sourcing

Contractors can be located anywhere in the world where labor is abundant; wages are reasonable; and facilities, machinery, and transportation are available. In spite of technological advances, apparel manufacturing remains labor intensive. Although garment workers in the

United States are paid relatively low wages compared with other skilled artisans, manufacturers have turned to even cheaper sources of labor in Asia, Eastern Europe, Mexico, and the Carribean. Apparel is manufactured in whatever country can produce the best garment for the lowest price. Often fabric is purchased in one country, exported to another country for sewing, and then shipped to a third country to sell. Liz Claiborne, for example, sources, or contracts out, approximately 90 percent of its requirements offshore with some 300 independent contractors in 40 nations.[1] According to the American Textile Manufacturers Institute, more than 61 percent of apparel puchased in the United States during 1992 was made either entirely offshore or of imported fabric.[2] Sourcing situations change every month, so many manufacturers find that they must spread out their production so that one country is never the predominant source of production.

Asia

Overseas production started in Hong Kong, which became the capital of Asian apparel manufacturing. Hong Kong companies became specialists in knitwear. Now those workers have become so skilled and well paid that only the highest quality clothing is made there. Manufacturers then sought cheaper labor sources in Taiwan, China, and South Korea. China is now the largest foreign apparel production supplier to the United States.[3] As those countries increased their skills and wages, the Association of Southeast Asian Nations (ASEAN), which includes Indonesia and

Workers at a small contractor in India - no modern technology here because labor is so cheap. (*Courtesy of Katrena Bothwell Meyer*)

Malaysia and other newly developed Asian economies including India and Sri Lanka, were found as new sources. Compared with an average hourly wage of $7.55 in the United States, typical hourly wages (in U.S. dollars) are as little as 25 cents per hour in China, Thailand, Pakistan, and India, and 15 cents an hour in Indonesia.[4]

Eastern Europe

Countries such as Poland, Hungary, and Slovakia in Eastern Europe, that used to primarily supply the former Soviet Union, are providing cheaper skilled labor for European manufacturers. Some American fashion companies such as Liz Claiborne and Levi Strauss are producing there as well. These countries have had to adjust from working in a command oriented economy, where delivery schedules were not a problem, to a market driven economy.

Mexico and the Caribbean Basin

As stores increasingly order closer to season, manufacturers are sourcing closer to home. Mexico is already the world's sixth largest apparel manufacturing country and under the North American Free Trade Agreement, production would continue to grow. The island countries of the Caribbean and Middle America who also enjoy tax and quota exceptions for apparel they produce for the United States hope that NAFTA will be widened to include them (NAFTA is discussed in Chapter 12). Sourcing for the U. S. is mainly in the Dominican Republic, Mexico, Costa Rica, Guatamala, Honduras, and Jamaica.[5]

Overseas Production Methods

Importing, bringing merchandise in from other countries, involves numerous negotiators: an agent to represent the manufacturer in the country where production is done, a customs broker to assist in processing the import application papers with the U.S. government to bring the goods into the United States, and a freight-forwarding agent to handle shipping. Manufacturers must rely on a good agent or company representative to keep control of quality, prevent late deliveries, and handle difficulties in communication caused by language differences. Price negotiations are usually made in U.S. dollars because of the fluctuations in international currency-exchange rates.

Manufacturers must allow approximately thirty days for sea transportation or three days for air delivery from Asia or Europe, plus trucking within the United States. An extra four days to two weeks must be allowed for customs clearance.

To produce overseas, manufacturers must send patterns, specifications, and samples as guidelines. All details must be exact and clear. Precise records and open lines of communication are a necessity. When production is done overseas, fabric is shipped directly to the contractor who does the cutting and sewing. There are three basic methods of producing clothing overseas: a production package; cut, make, and trim; and offshore assembly.

Production package. A manufacturer purchases a production package through an agent. In this case, everything originates in the production area, including raw materials, production, finishing, labeling, packaging, and shipping. Using an agent is the most expensive method, but it is advantageous because the agent takes responsibility for production, quality control, and the delivery schedule. A manufacturer may also have its own representative abroad to find raw materials, work with agents and contractors, and oversee production, quality control, quotas, duties, and shipping.

Cut, make, and trim. A manufacturer may buy fabric from one country—silk from China, for example—and then send it elsewhere to be cut, sewn, finished, and labeled.

Offshore assembly. Fabric is purchased and cut in the United States and sent to Mexico or the Caribbean countries for sewing. This classification, item *9802* in the Harmonized Tariff Schedule (still referred to as the *807 program*), is being encouraged to promote the use of American textiles. For this reason there is a *guaranteed access level* (large quotas) for this classification. Duty is charged only on the portion of labor done offshore.

Some American manufacturers have purchased or opened their own production facilities in Asia, the Caribbean, or Mexico. Asian firms are doing the same and investing in America and the Caribbean.

The pros and cons of sourcing abroad

The advantage of contracting overseas has been to keep production costs down. However, even though labor is inexpensive, obstacles such as customs duties (tariffs), quotas, freight costs, lengthy lead times, and maintaining production controls at a distance, or having executives move to Asia to supervise, are making it less advantageous than it formerly was. Also, labor costs in Asia are rising and there are many hidden costs involved such as travel to and communications with Asia. The trend toward increased overseas production continues, however.

Understandably, unions are fighting overseas production because it takes jobs away from American laborers. Many manufacturers would like to bring production back to the United States, especially to save lead time, but find that American textile mills are not flexible enough. The fabrics that U.S. mills produce is good but limited and the minimum orders are too high. So manufacturers seek both cheaper labor and more fabric choices abroad.

GARMENT ASSEMBLY

The next step in production is the actual assembly or sewing of quantitites of garments.

Assembly Operations

The steps involved in garment assembly are called *operations.* A man's suit can have as many as 200 different sewing operations. No two manufacturers use all of the same methods, but all of them follow the same basic order.

As part of production scheduling, a supervisor analyzes a garment's construction to determine the best and fastest way to sew the garment. An *operation sheet* is drawn up, listing all necessary operations in sequence. Specific sewing operations may be periodically reassigned because certain machines needed for one style may not be needed for another or may be altered to perform different functions for new styles. The introduction of specialized operations or new machinery requires additional training of operators that results in higher production costs during the training period. As apparel manufacturing remains labor intensive, manufacturers and contractors find it necessary to rely on their operators as individual skill centers or as part of a small production team.

Individual incentive systems (piecework)

Many sewing machine operators, finishers, and pressers are paid on a *piecework* rate. That is, they are paid a set amount for each operation that they complete, rather than by the hour. Rates vary with the difficulty of the operation. As proof of work completed, the operator signs the identification ticket or removes one segment of it. Actually, most companies pay a guaranteed wage, with piecework acting as an incentive for operators to work faster and therefore earn more.

Assembly systems

Sewing operations must be performed in sequence. There are two methods of construction: the progressive-bundle system and the tailor or whole-garment system.

Progressive-bundle systems, used most often for sportswear, require each operator to repeat one assembly task, such as closing a shoulder seam or stitching on a pocket. Referred to as *section work*, this method groups machine operators to follow the order of production. Garments are passed in units of forty or fifty from one section to the next as each operation is completed.

Modular manufacturing involving a group of people who work together to produce finished garments. *(Courtesy of the National Apparel Technology Center)*

The advantage of progressive-bundle systems is that the work area can be designed to complete a specific operation utilizing the best machinery and trained operators for that job. However, the pressure of fashion change makes long-line production less effective as it cannot respond to rapid change and there is a high work-in-process inventory.

Whole-garment systems use a single operator to sew a whole garment together. This system is similar to the one used for sewing sample garments. Finishing tasks may be performed by another operator who specializes in these operations.

Modular manufacturing is an innovative system which involves the cross-training of operators and development of small quality-circle production groups as used by the Japanese. Modular manufacturing involves a group of four to seventeen people who work together to produce a finished garment. Each team member works on more than one operation and group incentives may replace or supplement individual piecework.

Computers Used in Garment Assembly

To meet the competition of imports, factories are installing automated systems for movement of goods and sewing to speed up production and cut lead time.

Computer Aided Manufacturing (CAM) applications include computerized pattern making and grading, marker making, cutting, and programmable sewing machines.

Unit Production Systems (UPS) use conveyors to move garments automatically from one work station to the next where each operator is responsible for one stage of the assembly process (see progressive-bundle systems).

Computer Integrated Manufacturing (CIM) can connect these systems and share data for total automation. For example, a manufacturer can network data from a CAD patternmaking, grading, and marking system directly into a CAM cutting system by computer.

Computer simulation allows for the viewing of a manufacturing floor plan on a computer screen prior to actual implementation. Through simulation studies, the type and number of machines as well as the number of operators needed may be determined before the assembly process begins.

This unit production system allows operations to be performed without removing parts from the hanger, thereby reducing production time. *(Courtesy of Gerber Garment Technology, Inc.)*

Flexible manufacturing is simply a variety of strategies used by one manufacturer to produce apparel. Not all garments lend themselves to traditional or automated technologies; a flexible system uses a combination of methods.

Power Sewing Machines

Three main types of power sewing machines are used in traditional factories: the lock-stitch, the chain-stitch, and the overlock.

The lock-stitch machine sews a straight seam on the same principle as a home sewing machine. This machine makes it possible for the top thread to go under the bottom thread around a bobbin, creating a lock. This is the most secure stitch possible, but it leaves an unfinished seam, undesirable in fabrics that ravel easily. Also, of course, operations must be stopped frequently to rewind the bobbin.

The chain-stitch machine works on a principle similar to crocheting: it makes a series of loops pulled through one another. The top needle goes in and out of the fabric, making loops underneath that catch into one another. The chain-stitch is not as secure as the lock-stitch but, because the chain-stitch machine does not have a bobbin, the operator does not need to stop in mid-operation to rewind it.

The overlock machine creates a finished edge as it makes a seam. *(Photographed by the author)*

The overlock or serging machine is based on the same principle as the chain-stitch machine. It was created to make an edge finish as well as to sew seams. In one operation it sews the fabric together, cuts off the fabric to make a smooth edge, and wraps thread around the edge. A simple overlock machine has one needle and two loopers (which look like thick, bent needles), and works with three spools or cones of thread. The needle and loopers work together in a reciprocating pattern, the loopers moving back and forth from the needle to the fabric edge. This stitch is ideal for knits because it gives with the stretch of the fabric.

The safety overlock machine is a combination of the chain-stitch and the overlock. The safety factor is that in case one row of stitching comes out, the other still holds the garment together. With a total of two needles, three loopers, and five cones of thread, it functions as two machines in one. It provides the straight chain-stitch needed for factory assembling plus an edge finish.

The blind-stitch hemming machine is also based on the chain-stitch. The hem is folded back and caught by the needle at even intervals.

Button machines sew buttons onto a garment. Button placement is marked by tiny holes punched into the fabric. A sew-through button is placed in a holder, which moves the button back and forth while the needle sews it onto the fabric underneath. A shank button is held in position sideways so that the needle can go through the shank on its back. Some machines have feeding devices that automatically position the button.

The buttonhole machine is essentially a zigzag lock-stitch machine with automatic devices to control the width and length of the buttonhole and to cut it open. The operator must be careful to insert the fabric in exactly the right position each time, because the machine makes the buttonhole in about six seconds.

Programmable Automatic Sewing Machines

Automating sewing is more difficult than other factory tasks. The problem is handling the fabric as it is sewn. Moving silk through a sewing machine is very different from moving wool, for example. Generally, people must first load fabric pieces into the sewing machine. Automated sewing is done by a microprocessor-controlled sewing machine with an arm that can grasp the cloth and move it through the machine, align the fabric, and stitch it automatically. Sensors activate the sewing operations. However these machines can only do simple operations.

The clothing industry remains labor intensive for two reasons. First, textiles are floppy and soft, which makes it difficult for robots to handle them and for computers to simulate their shapes. Second, since the industry is fashion sensitive, automated machinery must be able to change as fashion changes.

Labeling

Along the production line, various labels must be attached to the garment. These include: manufacturer and/or designer identification, fiber content, care instructions, country of origin, size, and a union label if made in an American union shop. Apparel firms commissioned to manufacture clothes for stores that wish to sell garments under their own brand name require a *private label.* Federal law requires listing fibers in order of percentage used, along with cleaning or washing instructions.

A *country-of-origin label* is required by customs if the garment is made outside the United States. The label must say "made in (name of country)" or "made in USA of imported fabric." A "Made in USA" label is reserved for garments made entirely in the United States of American fabric. Items made partially in a foreign country and partially in the United States must disclose those facts; for example, a tag might read: "assembled and sewn in the USA of components made in (name of country)."

In addition, there may be a hang tag to promote the manufacturer or the fiber or fabric producer's name. Some manufacturers are now putting more information on hang tags to help the consumer. A manufacturer may also *pre-ticket* garments with UPC symbols for the retailer.

Finishing

Tailoring, the unseen quality handwork formerly done inside collars and lapels to form and hold their shape, has virtually disappeared because of the high cost of skilled labor. Most "tailoring" today is done simply by fusing interfacings into the garment to give it shape.

Hand finishing is usually seen only in couture garments. In better garments, some hand finishing is necessary, such as sewing in linings or sewing on buttons, but this too is becoming rare. In moderate-priced and lower-priced garments, all finishing is done by machine.

PRODUCTION OF MEN'S SUITS

The production of men's suits is one area where we still see tailoring.

Traditionally, the men's clothing industry has been divided into firms that produce tailored suits and coats, and furnishing producers that make shirts, slacks, ties, underwear, and sleepwear. Tailored garment production has dominated the industry, but with the trend toward casual dress for men, there has been a growing demand for sportswear. The production of men's sportswear is similar to that of women's.

Production of a man's suit, however, can require over 200 steps. Inner construction to give a suit shape and body makes production lengthy, complicated, and costly. Modern technology has introduced automation

A fitting at Savile Row tailor Gieves & Hawkes. *(Courtesy of Gieves & Hawkes)*

into an industry that was traditionally handwork oriented. Fusible interfacing has replaced hand-applied canvas in chest pieces, shoulders, lapels, collar, and buttoning areas to supply the required body and shape. Hand tailoring can be found in only the most expensive suits such as Oxxford in America or custom-made and other high-quality suits in Europe or Hong Kong.

PRODUCTION OF KNITWEAR

The production of knitwear requires special skills and machinery.

There are two commercial methods of knitwear production: cut-and-sew and full-fashioned.

Cut-and-sew is a method by which garments are made from knitted yardage. This type of knitted garment requires the same patterning as that used for woven fabrics, except that the amount of stretch in the fabric must be considered. Stretchability is the key to knitted garments; therefore an overlock stitch is used, because it is flexible and gives with the fabric. Knitted braids and bands must be used as trims and finishes because all components must have the same stretchability.

Full-fashioned knits, mostly sweaters, are actually shaped on the knitting machine. The specifications for each piece are programmed into the knitting machine, which forms the shape as it is knit by adding or dropping stitches at the edge of each piece to widen or narrow it. Then the pieces are joined together. Most full-fashioned production is done in the Orient because manufacturers saw the potential of cheaper labor there and invested in specialized machinery. Factories in the Orient provide both affordable prices and variety of production.

Computers have also changed the production time and cost of knitting. Computer-aided design (CAD) systems permit designers to see a pattern or garment design on the computer screen or printed out on paper without having to knit a sample. Knitting machines tied into the computerized design screens will accept pattern and stitch changes in minutes rather than hours, allowing greater freedom for experimentation. Adrienne Vittadini remarked, "With knits, you are also a textile designer; you need

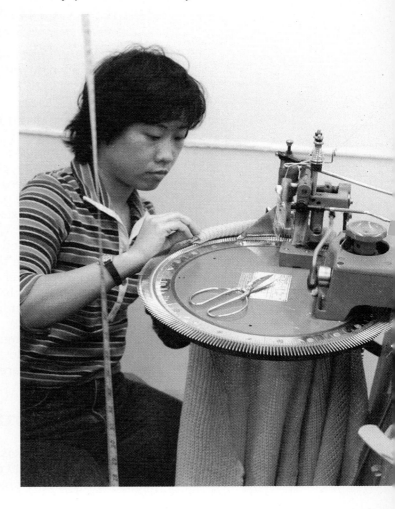

Joining sweater pieces on a linking machine at a knitting contractor in Hong Kong.
(Courtesy of Katrena Bothwell Meyer)

technical knowledge to achieve what you want and you must know how to use the machinery."[6] Each type of sweater, whether it is intarsia, jacquard, shaker, cable, or pointelle has a different lead time, quantity and weight of yarn required, gauge (stitch size), finishing treatment, and special skill and machinery needed to produce it.

Pattern is introduced by Jacquard and intarsia knitting systems. *Jacquard* patterns are created by various needle arrangements on electronic knitting machines. *Float Jacquard* shows the pattern on the face of the sweater while the yarns are carried on the back side until needed again for the pattern. To create a *full Jacquard*, both the front and back beds of the knitting machine are used, creating a heavier double knit. The main design appears on the front with another simple pattern on the back to avoid floats. For *intarsia* patterns, the yarn is knotted and cut off in back when no longer needed for the pattern on the front.

Some sweaters are also hand-knitted or crocheted. Hand knits have the greatest variety of stitches and detailing. However, because of the time needed to make them, they are very expensive. Hand-knit garments for commercial marketing are usually made abroad, where labor is cheaper and more readily available.

PREPARATIONS FOR SHIPPING

Pressing enhances a garment, quality control eliminates imperfections, and filling orders promptly facilitates on-time deliveries.

Pressing or Folding

Pressing vastly improves the look of a garment. It can hide a multitude of imperfections, such as puckered seams and collars that do not lie flat. Higher-priced garments are pressed during the course of construction, less expensive garments are pressed only after completion. Pressing equipment is sometimes used to shape pieces before sewing; for example, to fold under pocket edges.

Steam irons are used for areas not easily accessible. Various *buck* pressing machines, like those seen at a dry cleaner, are used on tailored garments to flatten jacket edges, crease trousers, and so on. Occasionally, a *steam-air form finisher,* which looks like a puffy dress form, is used to steam dresses into shape. Automated pressing is done by computer-controlled pressing equipment.

One of the popular trends in sportswear today, especially for denim, is to simply wash and fold the items. Environmentally friendly enzyme and stone washes have replaced acid washes.

Quality Control

A sample maker can make one garment neatly and accurately, but in the mass production of 500 or 1000 dozen garments it is harder to control quality. To make sure production has been done correctly and to prevent returns, both work-in-progress and finished garments are inspected either totally or by random sampling. Quality controllers not only check for poor sewing and uncut threads but also spot-check measurements against a list of specifications. If there are mistakes in first stock, the production manager tries to correct them at the sewing or cutting source.

Production standards are very important and must equal the manufacturer's reputation and warrant the garment's price. It means nothing to have a beautifully designed garment if it is not produced well. Inferior garments may be rejected by the store and returned to the manufacturer, or they may look so unappealing on the hanger that customers will not buy them. Either event would mean loss of reputation and future sales.

Garments returned to the manufacturer are analyzed to find the reasons for their return. Computer bar coding or numbers on each garment's label identify the factory or contractor where the garment was made, enabling construction mistakes to be traced to their source. Besides construction mistakes, reasons for returns include poor fabric and fabric shading. The manufacturer tries to prevent these mistakes in future shipments. Manufacturers have realized the need to improve the quality of their products as consumers become more demanding.

Environmental Concerns

Manufacturers of apparel are giving increased attention to environmental concerns. They are trying to identify ways to reduce and recycle their supplies. Their primary concern is the disposal of fabric scrap, which makes up a large percentage of production waste. One possibility for the disposal of this waste is to use an on-site incinerator to generate electricity. Levi Strauss has come up with an innovative method to recycle their scrap. They are using their denim scrap to make their company letter paper.

Many concerned manufacturers are also changing their packaging for shipping and no longer using polyester bags or styrofoam.

Filling Orders

If production work has been done by contractors, the completed garments are usually returned to the manufacturer for shipment to retail stores. At a central shipping point, garments are checked for quality and then divided into groups according to style, size, and color and put into *stock*, a storage area where they may be hung on racks or stacked on shelves.

A store orders a certain quantity of units by entering each style number, color number, and price per unit on an order form. The order is analyzed for a credit rating. The next step is to see what is available in stock to fill the order. This checking is now done much more easily and efficiently by computer automation. Orders can be transmitted by EDI and then automatically translated into invoices. Quick Response initiatives have paved the way for manufacturers to cut down the time that garments are held up in stock. Large orders and orders from important regular customers are filled first. Merchandise is pulled from stock to fill orders in the correct style, size, and color.

Garments are folded into shipping boxes and marked with the means of transportation specified by the retailer. A *packing slip* (an invoice without prices) is enclosed in the box, while the actual invoice is mailed separately for payment. The boxes are labeled with addresses and then moved by trolley or conveyor belt to be loaded on the delivery truck and sent to the retailer. On-time deliveries are extremely important to prevent cancellation of orders. On the basis of reliable production and delivery standards, a manufacturer builds and keeps a reputation for dependability that accompanies its reputation for innovative styling.

SUMMARY

In this chapter we discussed garment costing, pricing, and production scheduling. We have seen how garments are manufactured, from piece goods purchase and inspection, through patterning, grading, marking, cutting, assembly, and finishing, to quality control. Many manufacturers use domestic contractors to do all or part of their production and foreign contractors or off-shore production to reduce labor costs. Domestic factories try to compete with the newest computer technology, including inventory control, electronic data interchange and Quick Response strategies, as well as computer patternmaking, grading, marking, spreading, cutting, and assembly systems. Men's suits and knits present unique methods of garment production. Quality control and on-time deliveries help the manufacturer maintain a good reputation for dependability.

CHAPTER REVIEW

Terms and Concepts

Briefly identify and discuss the following terms and concepts:

1. Wholesale pricing
2. Cost sheet
3. Cost merchandising
4. Loss leader
5. Production schedule
6. Cut-to-order versus cut-to-stock
7. Piece goods
8. Contracting
9. Global sourcing
10. Electronic Data Interchange
11. Offshore assembly
12. Quick Response
13. Issue plan
14. Computer Integrated Manufacturing
15. Production pattern
16. Grading
17. Modular manufacturing
18. Unit Production Systems
19. Piecework
20. Lock-stitch machine
21. Tailoring
22. Cut-and-sew knits
23. Full-fashioned knits
24. Quality control

Questions for Review

1. Explain the costing of a garment.
2. What are some of the things to consider when purchasing piece goods?
3. Discuss the differences between the cut-to-order and the cut-to-stock methods of production.
4. Explain the difference between making and grading patterns by hand and by computer.
5. Describe the differences between three cutting methods.
6. Describe three ways in which computers aid apparel manufacturing.
7. Explain various systems of garment assembly.
8. Why do manufacturers use contractors?
9. Discuss the three major methods of overseas production.
10. How has the construction of men's suits changed over the years?
11. Explain the two methods of producing knit garments.
12. Why is quality control so important?

Projects for Additional Learning

1. If there is an apparel manufacturer or contractor in your area, try to make an appointment for a factory tour. Observe production methods. Do operators do section work, or does one person sew one whole garment together? What computer technology is used in the factory? What production methods are unique to the product? Compare the wholesale price range of the garments being produced with the quality of production.
2. Visit a local store and compare imported ready-to-wear with domestically produced items that retail at the same price. Consider fit, styling, and quality. Which is better? Why? What conclusions can you draw from this comparison?
3. Investigate a major apparel manufacturer by interviewing managers, reading the company's annual report, reading trade newspaper articles about the company, or writing to the company for information. Write a profile of the company including ownership (see Appendix Two), type of garments produced by each division, type of customer to which they appeal, type of production, price line, and where products are sold.

NOTES

[1] Nancy Marx Better, "The Secret of Liz Claiborne's Success," *Working Woman* (April 1992), p. 96.

[2] "ATMI Calls '92 a Better Year," *Women's Wear Daily*, December 30, 1992, p. 8.

[3] "U.S. Industrial Outlook 1993," U.S. Dept. of Commerce (January 1993), p. 32-7.

[4] "Textile Report," *Women's Wear Daily*, January 28, 1992, p. 27.

[5] "Industrial Outlook," p. 32-8.

[6] As quoted in Vittadini press release, p. 2.

Beatrice Amblard sews a Hermès bag. *(Courtesy of Hermès of Paris, photographed by the author)*

11

ACCESSORY AND FUR MANUFACTURING

CAREER FOCUS

Because this chapter covers so many categories, it includes many interesting areas of work. Although some companies produce several kinds of accessories, each area needs its own product developers, merchandisers, designers, pattern makers, production engineers, marketers, and sales representatives.

There are less opportunities in the fur industry as production is limited to a few cities in colder climates or where cheap labor has encouraged growth of the industry. However, marketing careers are possible in major fur centers.

CHAPTER OBJECTIVES

After reading this chapter you should have attained competence in the following areas:
1. The ability to discuss the unique design considerations of various accessories.
2. The ability to describe production methods for the major accessories.
3. Knowledge of accessory design and production centers.
4. The ability to explain fur garment production.

This chapter discusses manufacturing as it particularly applies to accessories and furs. Handbags, shoes, belts, gloves, and furs have in common the age-old use of animal skins as a raw material. There are separate sections covering the styling and production of shoes, handbags, belts, gloves, hats, scarves, hosiery, jewelry, and furs because each type of accessory has its individual methods for design and production. Although fur coats and jackets are wearing apparel, we discuss them separately from other garments because of the specialized methods used in their manufacture.

FASHION ACCESSORIES

Accessories such as shoes, handbags, belts, hats, and jewelry are designed to coordinate with apparel to create a total fashion look.

Over the last several years the accessories industry has become one of the most exciting segments of manufacturing and retailing. Accessories give consumers a way to update their wardrobes. The popularity of various accessories is cyclical. Footwear and handbags are always needed but the fashion for belts is related to waist interest in apparel. Accessories such as jewelry and scarves become more important when clothing is classic and simply styled. Fashion in hats and hair goods is obviously related to hair styles. Designers and producers of accessories must be aware of fashion trends in order to make accessories that successfully complement apparel.

To balance product lines in the face of fashion cycles, many companies are diversifying into other areas. Echo (scarves) has added a belt line; Ferragamo (shoes) now makes handbags, jewelry, and even ready-to-wear.

Hermès has added ready-to-wear to their famous line of scarves and leather goods. These firms try to carry a design element throughout their accessory lines. A chain belt might repeat the design of a chain strap on a handbag or a handbag clasp might be copied as a belt buckle. Judith Leiber, for example, has used an animal clasp on a belt to go with an animal design handbag.

Since many apparel designers want accessories that are specifically created to go with their apparel collections, leading apparel companies such as Liz Claiborne and Donna Karan have added accessories to their product lines through licensing arrangements or through joint ventures with accessory manufacturers. Retailers, too, are trying new strategies to sell more accessories. Many stores have opened accessories shops on their ready-to-wear floors, whereas traditionally, retailers have had accessory departments only on the main floor. This new arrangement helps sell accessories to complement apparel purchases..

Footwear

Footwear, including shoes, sandals, and boots, is the largest volume producer in accessories. More than 7 billion pairs of shoes are produced worldwide each year.

Both functional and fashionable, shoes come in assorted materials including calf, kid, suede, and reptile skins, and, due to rising cost, imitations. Dressy and casual shoes for women may also be made of fabric, made popular by ethnic styling. Sport shoes have added nylon to the list of possible materials.

As a result of the enormous popularity of sport shoes, comfort has also become an important element. A number of popular shoe brands, from Florsheim to Ferragamo, have tried to combine style with the comfort of athletic shoes.

Design of leather fashion shoes

Most fashion shoe design direction comes from Europe. Creative international shoe designers such as Maud Frizon and Andrea Pfister set international trends for women's fashion shoes. Apparel designers such as Gianfranco Ferré have shoe collections made by Italian shoe manufacturers, in this case Diego Della Valle, in joint ventures or via licensing arrangements. There are also some well-known American designers such as Kenneth Cole. Shoe designers study fashion trends so that their shoes will coordinate with apparel. They must also consider their materials and the view of the shoe from all angles.

Each company's *line builder* or product manager attends the shoe fairs in Dusseldorf, Germany, and Bologna and Milan, Italy, to get ideas for a new shoe collection. Like an apparel merchandiser, the line builder begins with concepts for groups and works with designers who develop individual shoe styles. Sometimes the line builder will buy *prototypes* (sample shoes) from a *modelista* (model maker) at a studio or shoe fair, otherwise the line builder might forward a designer's sketches to a modelista, who makes the first model at or near the factory. If the line builder and modelista live in different countries, ideas and samples must be sent back and forth. The sample shoes are edited to form a balanced collection.

Leather sourcing

Leather sourcing is usually close to production: Italian leather for shoes produced in Italy, South American leather for Brazilian-made shoes, and Chinese leather for Asian-made shoes. However, leather is sometimes sourced in one country, sent to another for production, and marketed in a third country.

Production of traditional leather shoes

Traditional shoe production is complex, involving various measurements for length and width combinations. The full range of women's shoe sizes includes 103 width and length combinations between sizes 5 and 10. Shoes for the American market are made on American lasts. European shoes have only one width whereas shoes exported to the United States usually come in four widths.

Another factor contributing to the complexity of shoe production is the skilled labor needed to complete the number of operations performed. Two to three hundred operations can go into the production of a finely made fashion shoe. If a shoe is factory-made, eighty different machines could be involved in its production.Whether a shoe is handmade or mass-produced, there are usually ten basic steps:

1. **Making the last.** A foot-shaped form called a *last* is created. The original last is made of wood and requires as many as thirty-five different measurements. A shoe factory needs thousands of lasts, one for each size, width, heel height, and basic style. Duplicate polyethylene lasts are made from the master.

2. **Patternmaking.** A pattern is created based on measurements taken from the last and from the original model, or designer's sketch. A

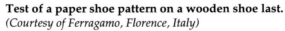

Test of a paper shoe pattern on a wooden shoe last.
(Courtesy of Ferragamo, Florence, Italy)

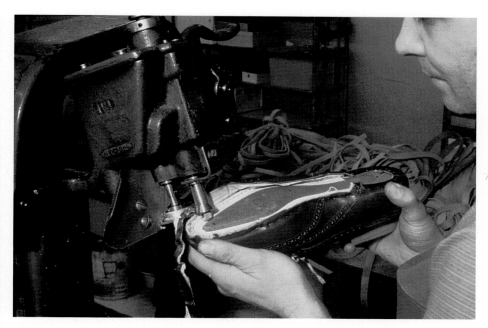

Attaching the leather shoe upper to the last. *(Courtesy of Bally of Switzerland)*

trial shoe is made from this pattern. The line builder further assesses and refines style selections at this time, and the final line is closed or "frozen." Trials are duplicated as samples for sales representatives.

3. **Cutting.** Paper patterns are converted into steel dies that cut the leather. Some ultramodern factories use water jets or laser beams to cut the leather.

4. **Stitching and fitting.** The upper portion of the shoe is made. Buttonholes, topstitching, all other upper design details, and linings are added.

5. **Lasting.** The innersole and the upper are attached to the last.

6. **Stock fitting.** The sole section is prepared.

7. **Bottoming.** The sole is attached to the upper by sewing, cementing, nailing, or molding.

8. **Heeling.** The heel is attached and shaped.

9. **Finishing.** The finished shoes are polished, the last is removed, and the brand name is added to the innersole.

10. **Treeing.** Decoration is attached followed by a final cleaning and inspection.

Large companies have a competitive edge because they have the capital necessary to invest in the most advanced machinery and computer technology. Shoe companies are using computer-aided-design (CAD) systems that are capable of both two-dimensional

Trimming the shoe lining at the Lerre factory in Naples, Italy. *(Courtesy of Lerre and Bucci Imports)*

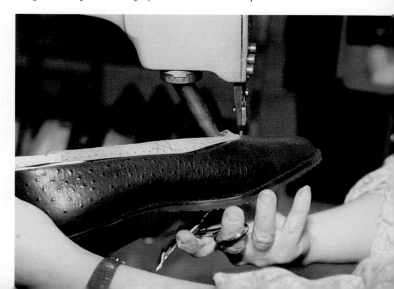

design (design of uppers and size grading) and three-dimensional design (design of the last and projection of drawing on the last). Programmed *continuous cutting machines* and *water-jet cutters* can be used for leather and synthetic materials. Widespread use of *computer-aided shoemaking systems* enable manufacturers to speed production, improve quality and keep labor costs down. Programmable machines are now available for nearly every aspect of production including sewing, folding, and lasting.

Imports

The greatest impact on the American shoe industry is the increasing number of imports. Shoe imports have reached a new high of 88 percent of the 1.1 billion pairs purchased annually in the United States.[1] Domestic shoe production has declined so that there are only 379 manufacturers operating 471 plants in the United States. The small amount of shoe production left, primarily of men's shoes, is done in New England and the Midwest. These firms include better classic men's shoes such as Johnston & Murphy, Allen-Edmunds, and Alden, and moderate shoe lines such as Florsheim, Nunn-Bush, and Bostonians. In the future, it seems that weaker firms will continue to close while stronger ones will consolidate plants and invest in new technology to narrow the gap between foreign and U. S. labor costs.

The highest quality women's fashion shoes, such as Andrea Pfister, Bruno Magli, Carlos Falchi and Ferragamo, are made by hand in Italy with the finest leathers. Some manufacturers of women's shoes such as Magli and Ferragamo also make men's shoes. Companies such as Borri and Lorenzo Banfi make men's shoes exclusively.

Bridge lines such as Anne Klein II, Via Spiga, Amalfi, or DKNY may be made in Italy, Spain, or possibly Eastern Europe. Better shoes including Liz Claiborne, Nickels, Caressa, or Pappagallo are typically made in Brazil.

Moderate, budget, and sport shoes are usually made in China, Korea, or Malaysia. Manufacturers are always seeking new cheaper production sources. Some companies buy production time in overseas factories. Others own facilities in Italy, Spain, or South America.

Most shoe importers, such as Schwartz & Benjamin, are essentially *marketers*. They buy or license the designs and contract production. They operate this way because they feel they have the knowledge of what products their customers want and the ability to market the shoes in their own country. There are companies that market shoe lines in all price ranges and categories. For example, U. S. Shoe markets Amalfi, Bandolino, Cobbie, David Evins, Easy Spirit, Evan-Picone, Joyce, Pappagallo, Selby and Vittorio Ricci. Many shoe manufacturers are vertical operations and own their own stores.

Markets

The prestigious international shoe fair, MICAM is held in Bologna in March to show fall styles and in Milan in September to show spring styles. There are also the GDS footwear show in Düsseldorf, Germany; MIDEC in Paris, France; and FICC in Elda, Spain. The Fashion Footwear Association of New York (FFANY) and the National Shoe Fair (for volume and lower-priced shoes) are held four times a year in New York. Buyers shop for early spring styles in June, for spring and summer in August, for early fall in December, and for fall and holiday in February. Shoe manufacturers use educational tools such as seminars to educate sales personnel as well as consumers in regards to quality and styling.

Handbags

A handbag must be both decorative and functional; it must hold necessities conveniently as well as fit into the fashion picture. Large bags such as totes, satchels, or portfolios tend to be more functional; smaller bags such as clutches or envelopes are usually decorative. Handbag styles range from classic constructed types to soft shapes. Leather, including suede and reptile, still represents approximately half of handbag material (60 percent of domestically produced bags but only 26 percent of imports); vinyl, fabric (tapestry, rug prints, needlepoint, metallics, nylon, and canvas), and straw make up the other half.

Design

Handbag design is very much like apparel design. The elements of color, line, shape, and fabric, as well as the principles of emphasis and proportion must be considered. To test a design, a *dummy* is made in flannel from a sample pattern. Ornaments, closures, and/or handles must be chosen to compliment the shape and fabrication When fabrication is selected, a final sample is made up with supportive interlinings that differ with each type of bag and each fabrication. Many design samples are made and then grouped around a theme. Several types of groups are needed to create a well-rounded collection.

Production

When the final line is chosen, cutting dies are made from the pattern and used to stamp out leather. The leather may also be cut by water jet and luxury handbags are cut by hand. Rising prices and reduced availability of leather in recent years have had a great impact on the styling and production of handbags. Fabric bags are cut by methods similar to those used in the apparel industry.

Luxury handbags.

The type and quality of workmanship varies greatly. At the top of the luxury market are Hermès handbags that are entirely handcrafted. Production is limited and is allocated by the number of hours it takes to make a bag. For example, it takes sixteen hours to make a Kelly handbag which costs over three thousand dollars. Each bag is dated and stamped with the craftsperson's initials. Hermès bags are made at their factory in Paris. Distribution is limited in order to maintain exclusivity.

Chanel handbags are in the $600 to $1000 range. They have two factories in France and one in Italy where the bags are hand molded and also limited in production. Kleinberg-Sherrill and Judith Leiber are the only American luxury handbag companies which actually produce in the United States. Most other luxury handbags are made in France or Italy including Bottega Veneta, Donna Karan, Paloma Picasso, Prada, Fendi, Ferragamo, and Calvin Klein.

Better handbags.

In the production of better or moderate handbags, both fabrics and leathers may be stitched by machine, but much of the assembly of linings, ornaments, handles, and closures must still be done by hand. Although this handwork is expensive, some moderate to better handbags are still made in the United States, primarily in New York, Maine, Connecticut, Massachusetts and Florida. However, to keep costs down, many manufacturers source worldwide for the cheapest labor. Of the imports, better handbags are made in Hong Kong. Other handbag production sources include South America, Indonesia, China and India.

Better or "bridge" handbag vendors include Bally of Switzerland, Kenneth Cole, Stone Mountain; American designer collections examples are Anne Klein or Donna Karan New York (DKNY); and classic lines include Coach, Etienne Aigner or Donney & Bourke.

Moderate and inexpensive handbags.

Moderate bags are sometimes made of leather but usually of vinyl or fabric. To keep prices down, manufacturers have turned to cheaper labor sources overseas. Handbag imports have risen to 90 percent of U.S. consumption and of these imports 76 percent are from China.[2]

Some moderate brand examples are Liz Claiborne and Esprit. To keep prices down, many stores create *private label* handbags in all categories. The retailer works directly with a manufacturer, labels the handbags with the store's private brand, and eliminates the cost of wholesale marketing.

Belts

The fashion for belts in women's wear is *apparel driven* (corresponding to waist interest in apparel styling). Rising prices and reduced availability of leather has also had a great impact on the belt market. Therefore, fashion belts are made not only of light-weight leathers and suedes, but also of leather look-alikes, fabric, and elasticized fabrics. Men's belts are strictly functional and must withstand everyday wear, so they are traditionally made of heavy five-ounce or six-ounce leather.

Many apparel companies such as Chanel, Donna Karan, Isaac Mizrahi, and Liz Claiborne have belts to accessorize with their clothing. Accessories companies such as Hermès and Gucci also make belts to coordinate with their handbags.

Production

Leather and imitation leather materials are cut on *strap-cutting* machines or on computer-aided cutting machines. These machines cut the material into long, straight lengths of any desired width. Shaped belts are made either by die-cutting or from Plexiglas patterns. With die-cutting, the pattern is made into a die resembling a cookie cutter. A *clicker* machine presses the sharp edges of the die through the leather. Plexiglas patterns are used like paper patterns for single-layer cutting. The cutter must carefully cut around the patterns, using only flawless pieces of leather. Computer-automated cutting has helped to speed up production and insure accuracy.

The cut leather is next sewn to a backing with a *walking-foot machine* or on computer automated machines. Belt buckles, made from metal, wood, or plastic, come mostly from New England and Italy. Holes are made at one end of the belt by a *foot-press machine.* The holes, the slit for the buckle, and the shape of the tip may also be cut by a die. In addition, a die can emboss a pattern onto the leather. Belts are finished with trims such as stitchery, cording, nail heads, or rhinestones and may have an edge finish made by an edge die.

Industry organization

Belt manufacturers tend to be small firms. The industry is centered in the New York metropolitan area because of that city's proximity to suppliers and production is also growing in California. A large investment in equipment is needed to finish leather well, but some of the operations can be contracted out.

Imports

As in the manufacture of other accessories, belt production has moved to a large extent overseas. Many belts are now made in China, Taiwan, Korea and Hong Kong.

Gloves

The demand for gloves has declined over the years but there is some renewed interest. Although gloves are functional, they are designed to complement fashion suits and coats. Black remains the most popular color for gloves but brights are also included in glove collections. Glove fabrications are primarily leathers, cotton, and wool or acrylic knits. They are often trimmed with real and fake fur, buttons, bows, braid, embroidery, lace, tassels, and zippers.

Production

Most gloves are still produced by a painstaking process that requires many hand operations. Because of this, leather glove producers tend to specialize in a single step of production, such as cutting, stitching, or finishing. However, fabric glove production is becoming more automated.

Imports

Glove firms have remained small because of their specialized, handcrafted operations. Italy has been a traditional leader in the glove industry. Some Italian glove companies produce their own lines as well as gloves for major designers. Portolano, for example, also produces for Moschino, Fendi and Barry Kieselstein Cord. Many European and American glove manufacturers have either purchased plants or use contractors in China, the Philippines, or Eastern Europe, but even imports have declined due to the low demand for gloves.

Resources

Major glove manufacturers include Aris (Isotoner), Carolina Amato, Fownes, Grandoe, La Crasia, Portolano, and Shalimar. Some manufacturers are starting in-store training to educate salespeople in order to promote sales.

Hats

In the past, the most important accessory was a hat. A woman bought a new hat to add a bright spot to her wardrobe; a businessman was never seen on the street without one. The trend toward more casual life styles and bouffant hairdos of the 60s changed that, and the millinery industry suffered a severe setback. Of course, functional hats to protect against the weather remained a necessity.

Today, there is a new demand for hats on the part of young people. Fashion hats update or add pizazz to apparel for weddings and other special occasions. There is also a growing interest in soft hats and caps, to shield the sun's rays, which are worn for many recreational activities.

Production

Traditionally, fine hats were referred to as *millinery*. Designer millinery is hand-blocked over wooden forms and made of the finest materials. They may be decorated with ribbons, flowers, feathers, or other trims. Hats in the inexpensive to moderate price range are machine-blocked; crushable fabric hats are cut and sewn. The handcraft involved in millinery production tends to keep the industry small. There are only about one hundred hat manufacturers in the United States, located primarily in New York City and the Los Angeles area.

Resources

Successful hat designers and manufacturers include David (Cohen) Inc., Lola Ehrlich, P. J. Gedney for Hat Attack, Deborah Harper, Melinda Hodges, Eric Javits, Kokin, Frank Olive, Patricia Underwood, and Whittall & Shon. Most of the hat manufacturers also have less expensive lines. Leading apparel designers such as Donna Karan, Ralph Lauren, and

Hat designer Frank Olive works on one of his creations.
(Courtesy of Frank Olive, photographed by the author)

Calvin Klein also have hat lines to complement their clothing. These are usually made for them by hat manufacturers as joint ventures or under licensing agreements. For example, Hat Attack produces for Ralph Lauren. Other hat manufacturers include Betmar, Maeve Carr, Commodore Corp., Kerry Kahn, Makins, Mocha, and Deborah Rhodes. World centers include Paris, Milan, Germany, England, and Australia.

Marketing

Stores are opening hat shops on apparel floors to help customers accessorize their clothing purchases. Hat designers are using sales training and personal appearances in the stores to promote their designs. To promote millinery, many upscale stores create comfortable areas with chairs, mirrors, and special lighting. The Millinery Information Bureau gives annual "Milli" awards for outstanding hat design.

Blocking a hat form by machine in a factory.
(Photographed by the author at Sonni Hats)

Hair Accessories

The hair accessory business has grown tremendously in recent years. Of course it is directly related to hair styles and fashion trends. There are many categories including bows, barrettes, head bands, ponytail holders, and combs.

There are thousands of little resources for hair accessories. It is such a fast-turn business that manufacturers have to rely on local production. Due to the handwork involved, *cottage industry* (home work) is used to make them. Popular resources include Colette Malouf, Anne Vuille, David, Kirk's Folly, and the largest—Riviera.

Scarves

The popularity of scarves, shawls, and stoles, like other accessories, runs in cycles. Consumers want variety so it is necessary for a company to produce many designs. Design is usually based on historic textiles and layouts are painted to fit the desired shape.

Scarves can be made of silk, wool, cashmere, cotton, and man-made fibers. The fabrications include challis, sheers, metallics, and knits that can be printed, embroidered, or beaded.

Production

Scarf production is similar to textile production. The difference is in the printing and finishing. High-end scarves, those made of the finest fabrics, are made in Italy. Most of the printing is done by silk screen, which lends itself to the square or rectangular shape of the print. Hermès, which make the finest

16 to 50 color screens are used to make a Ferragamo scarf. *(Courtesy of Ferragamo)*

scarves, uses an average of 24 color screens and up to 50 screens for their highest quality! Successful moderate resources include Echo and Collection XIIX. Less expensive scarves use only 4 to 10 color screens. The printed fabric is cut and then hand-rolled, machine-rolled, flat-hemmed, or fringed.

Most scarf production is in Asia where expertise as well as inexpensive labor can be found. Many American manufacturers buy their fabrics in Korea or China and have the scarves printed there as well. India is a resource for novelty scarves.

The high price of silk, the cost of labor, and the demand for scarves has caused rising prices. Retailers have to order early to get on-time deliveries because of the long lead time involved in overseas production. Scarf manufacturers have developed styling videos and brochures to help the customer and promote sales.

Hosiery

Panty hose, stockings, and socks serve the practical function of keeping legs and feet warm. Women's hosiery also makes legs look attractive. The fashion for pattern or color on the legs is cyclical and does not necessarily coincide with variations in skirt length. Decorative hosiery was fashionable with the miniskirt of the mid 1960s as well as with the longer skirt lengths of the 1980s. The development of panty hose in the 1960s substantially added to the growth of the hosiery industry. The popularity of fashion socks as a sportswear accessory is a growing part of the industry.

Hosiery is produced in knitting mills where machines run 24 hours a day. Most hosiery mills perform all steps in production, although some smaller mills contract out the finishing processes. Full-fashioned hosiery is shaped as it is knitted, fulfilling size and length specifications. Panty hose are knit on circular knitting machines; stockings are flat-knit and then seamed. The steps involved in hosiery production are knitting, dyeing, boarding (a heat-setting process), pairing (for stockings), brand stamping or labeling, and packaging.

There are approximately 800 hosiery mills in the United States, located primarily in North Carolina, other southern states, and Pennsylvania. The second largest accessories industry, it is dominated by large companies such as Hanes, Round the Clock, Berkshire, and Hampshire.

Stock replenishment is especially important in the hosiery business. Manufacturers such as Hanes are using Quick Response systems such as electronic data interchange (EDI) and bar coding to keep track of and replace inventory. An interesting recent development is the growth of retail chains specializing in hosiery and/or socks.

Jewelry

Fine jewelry

Fine jewelry is costly because of the precious materials and the craftsmanship that go into it. Because fine jewelry is often a life long investment, its design is usually classic. However, there is currently a trend in high-style contemporary design led by designers such as Elsa Peretti, Paloma Picasso, Angela Cummings, and Robert Lee Morris.

Jewelry designer Robert Lee Morris at work in his studio.
(Courtesy of Robert Lee Morris)

Only precious metals such as gold, platinum, and some silver are used to make fine jewelry. Recent fluctuations in metal costs have made pricing difficult for the jewelry industry. Since precious metals in their pure state are too soft to retain shape or to hold stones securely, they are combined with other metals. Gold content is expressed in *karats*. Twenty-four karats is pure gold; 14 karats is 58.3 percent gold. *Goldsmiths* use these precious metals to create jewelry and make settings for precious gems. New York, Italy, and France are international centers for the creation of fine goldwork.

Precious gems are hard natural stones selected for their beauty. *Gemstones* include diamonds, rubies, emeralds, sapphires, alexandrites, aquamarines, topazes, tourmalines, garnets, jade, opals, lapis lazuli, and turquoise. Price depends on clarity, color, rarity, and size. The weight of gemstones is measured in carats, a standard unit of 200 milligrams. The term *carat* comes from the seeds of the carob tree that were once used to balance the scales used for weighing gems.

Diamonds have traditionally been the most valuable and coveted of gems, as Lorelei Lee sang in "Diamonds Are a Girl's Best Friend." Diamonds are the strongest natural element known: a diamond can be cut only with another diamond. Eighty-five percent of the world's diamond production is controlled by DeBeers, a huge South African conglomerate.

Transparent stones such as diamonds and aquamarines are cut by a *lapidary* (stonecutter) into symmetrical facets to show off their beauty. At least 50 percent of a rough gem is wasted in cutting. *Cabachon* stones—the unclear stones such as jade, opal, and coral—are domed, carved, or left in their natural state. Cabachon rubies and sapphires are also treated this way, a process that results in star rubies and sapphires. Although computers have recently been programmed to direct the cutting of gemstones, it is by and large still a craftsperson's field. Major stonecutting centers are in Antwerp, Belgium; Tel Aviv, Israel; London; New York City; and Idar—Oberstein, Germany.

Costume jewelry

Bridge jewelry is a category of better costume jewelry between fine jewelry and less expensive costume jewelry. Bridge jewelry is either sterling silver, gold filled or gold electroplated.

> *Vermeil* is a process of electroplating gold over silver (a new meaning given to the name of an old hand-rubbing process). *Electroplating* is a method of electric deposition that coats the metal piece with gold.

> *Gold-filled jewelry* is made from a thin metal sheet, usually brass, that is sandwiched (mechanically bonded) with a very thin film of gold on either side. When shapes are cut or stamped out of the sheet, the edges must be covered with gold.

Traditional costume jewelry simulates fine jewelry, using base metals such as brass, aluminum, copper, tin, lead, or chromium, coated or bonded with gold or silver.

Fashion jewelry utilizes metals that imitate gold and silver as well as materials such as wood, plastics, leather, beads, glass, or clay. Fashion jewelry is often colorful and styles change seasonally.

Jewelry production

There are as many techniques of making jewelry as there are jewelry types. Artisans use various methods of carving, grinding, drilling, filing, hammering, and welding to achieve the desired shapes. There are also many ways to mass-produce jewelry by using sheet metal, metal cast in molds, and wire.

Flat shapes are usually stamped out of sheet metal and may be decorated with embossing or engraving.

Casting is a process used to produce three-dimensional shapes. Rubber molds are used to cast low-temperature metals such as tin alloys; *lost-wax casting* is used for high-temperature metals, such as gold, silver, and brass (for both fine and costume jewelry). In the latter method, wax is first formed in the rubber mold and a new plaster mold is made over the wax forms. The wax is then burned out and molten metal is forced into the plaster mold. Finally, the plaster mold is broken open to expose the shaped pieces of metal, which are then snipped off of a supporting tree structure and polished.

Wire is used to make chains and various necklaces and bracelets. Band rings may be made by slicing tubes of metal. Fake jewels and pearls are made from glass.

Designs may be applied with enamel. Enamel work is distinguished by the way it is applied; methods include cloisonné, champlevé, basse taille, and painting.

There are approximately 750 costume jewelry firms in the United States. Production is located predominantly in New England, especially around Providence, Rhode Island, and the Middle Atlantic states.

Resources and marketing

Chanel and Dominique Aurientis are examples of companies that make fine costume jewelry. Fashion jewelry resources include Robert Lee Morris, Christian Dior, Anne Klein, Jay Strongwater, Carolee, R. J. Graziano, Medici, Trifari, Napier, 1928, Swarovski, Ciner, Art Deco, Miriam Haskell, Rhea, and Erwin Pearl.

Large firms like Crystal Brands (incorporating Monet, Trifari, and Marvella), Napier, Carol Dauplaise, and 1928 are among jewelry vendors that are emphasizing consumer and sales staff education to promote multiple sales. Trifari, for example, has full-time fashion accessory consultants on the selling floor who are paid in part by Trifari and in part by the store. They also have a series of booklets to show the consumer how to accessorize with jewelry and what styles look best with particular necklines. One of the booklets has punch-out photographs of the jewelry so that a customer can "try it on" at home. Another trend in the business is specialty retail stores devoted solely to costume jewelry.

Watches

As watches have become fashion accessories, luxury watches have become status symbols. They have intricately detailed mechanical works, usually hand-wound or self-wound, and some are quartz powered. Many include alternative functions such as the phase of the moon, the corresponding time half way round the world, chronograph capacity which indicates time intervals, perpetual calendars, and/or an accumulator that indicates elapsed time, shock resistance, and water resistance. Switzerland produces 95 percent of the world's luxury watches.

Many luxury watches are 18—karat gold and some are covered with diamonds. The most expensive brands—Breguet, Blancpain, and Patek Philippe—can cost up to a million dollars. Other top-end luxury watches include Audemars Piguet, Gerald Genta, Ulysse Nardin, and Vacheron & Constantin. Less expensive luxury watches are Rolex, Ebel, Cartier, Baume & Mercier, Corum, Piaget, and Omega.

In 1967, a Swiss firm produced the first quartz watch which was copied by Japanese and American fashion watch companies. The Swiss also developed the Swatch which set a direction for international fashion watches at popular prices. Upscale watch brands include Movado, Tissot, Raymond Weil, Rado, Fendi, and Gucci. Popular fashion watches include Swatch, Fossil, and Guess. Most fashion watches now are battery-activated quartz crystal, computer chip, or electronic motor.

FUR MANUFACTURING

Furs share the raw material of animal skins with leather accessories and were the earliest form of clothing.

In Chapter 7 we discussed the processing of furs. Some manufacturers buy furs directly at auctions and contract out the processing operations, or have their own vertical operations. Other manufacturers buy processed furs from merchants. Once the pelts have been processed, the actual production of fur garments can begin.

Styling trends

Fashion trends influence the design of fur garments and the popularity of specific furs. One current trend is reversible coats with one side fur and the reverse side treated like leather. Another trend is to trim textile coats, suits, or jackets with wild furs or combine leather with fur. This is causing a blurring of the boundaries between textile and fur garments and attracting new, younger customers.

Industry Organization

Fur garment styling centers include Milan, Paris, Frankfurt, Hong Kong, Montreal, and New York. Famous fashion names in fur manufacturing include Alixandre, Ben Kahn, Christensen, Corniche, Danzl, Grosvenor, Fendi, Maximilian, Revillon, Solecitti, Teso, Yves St. Laurent, and Zuki. Some well-known apparel designers, such as Ferré for Mondialpelli or Montana, Christian Dior and Donna Karan for Christensen of Copenhagen, create styles for fur manufacturers on licensing arrangements.

A sheared beaver coat designed by Zuki of Canada.
(Courtesy of Fur Council of Canada)

Only a very small percentage of coats are still produced in New York. The old-fashioned family furriers have virtually disappeared. The area of New York City between Sixth and Eighth Avenues and Twenty-sixth and Thirtieth Streets is the center for the creation of luxury fur coats in the United States. Furs designed by Valentino and Karl Lagerfeld are manufactured in New York.

On the world market, Hong Kong is the largest producer of fur garments, followed by Canada, Greece, and China.[3] Imports from Greece and China tend to be low end (low priced), must be purchased in volume, and ordered further ahead of selling season.

Production

No two fur pelts are totally alike; the furrier must match pelts according to quality and color to achieve uniform texture and color in the finished garment. The sewing of furs requires much skill and does not lend itself to mass-production techniques.

The two basic methods of fur manufacturing are the skin-on-skin technique and the letting-out technique.

Skin-on-skin method

In the less costly skin-on-skin method, one full skin is sewn next to another in a uniform alignment. This method is often used to sew together the leftovers of expensive furs, such as paws and flanks, into less expensive garments.

Letting-ot method

Luxurious furs, such as mink, are often manufactured by the letting-out method, which accentuates length, reduces width, and enhances draping. This technique involves splitting each skin in half lengthwise and then slicing every half pelt into diagonal strips one-eighth to one-quarter inch wide. The strips are then rematched and sewn together to form a narrower, longer skin that can run the full length of the garment. The result is a slimmer, longer pelt that is often more beautiful than the original. A let-out coat might have anywhere from 1000 to 20,000 seams which is one reason fur coats are so expensive.

Sewing

The strips are sewn into sections according to the coat or jacket pattern. They are first dampened, stretched, and stapled onto the pattern on a wooden board. After the sections are dried into shape they are sewn together.

The next operation is called *glazing*. The fur is again dampened and the hairs are combed in the desired direction. Gums and other materials, which often increase the luster of the fur, are applied to hold the hair in the desired position. Then the fur is slowly dried and the lining is sewn in.

Fur Labeling

Fur garment manufacturers may alter the appearance of various furs to resemble those that are currently more stylish or more expensive. Many countries have introduced labeling requirements to make consumers aware of this practice. The United States Fur Labeling Act requires that the label (as well as related advertising) contain the following information: animal name, country of origin, type of processing and dyeing, whether the furs have been reused, and whether the garment contains paws or tails. If the garment resembles another fur, it must be labeled with the name of the actual pelt used.

Stapling damp sections of a coat into shape. *(Courtesy of the Deutsches Pelz Institute)*

Markets

Annual fur fairs are held to show retail buyers fur collections from manufacturers around the world. International fairs include: the Tokyo Fur and Fashion Fair in February, Fur Industries International Salon (SIIF) held in Paris in March, Milan's Comispel in March, and Fur & Fashion Frankfurt in April.

The demand for furs depends on a number of variables: climate, the world supply of mink, economic conditions, and sympathies with animal rights activists. The United States is the largest market for furs followed by Japan. Once profitable markets in Britain, Germany, and the Netherlands have fallen off. Harrod's of London has closed its fur salon. A symbolic blow to the Canadian industry was the closing of the fur salons of the Hudson Bay department stores, a company which was built around the fur trade in Canada.

SUMMARY

Fashion in accessories is cyclical and related to apparel styling. Many apparel designers create accessories to complement their clothing. In the shoe business, a modelista makes the first models of shoe designs and then the line builder organizes them into a balanced collection. The production of shoes is a complex process from last making to finishing. Manufacturers are using computer-aided machinery to speed up production and save labor. In handbag production, dummies are first made to test a design. Expensive bags are made by hand of quality leather while less expensive bags are made of imitations or fabrics and are imported. The manufacture of belts is done at many small companies. Glove production involves many hand operations; most gloves are imported. Designer hats are hand-blocked and made of the finest materials while less expensive fabric hats may be cut and sewn. High-end scarves are made in Italy with many color screens on the finest fabrics while less expensive scarves are made in Asia. Hosiery manufacturing is dominated by large knitting mills. Fine jewelry is made of precious metals and gemstones and tends to be classic in design. Better costume jewelry is vermeil or gold filled and is made out of sheet metal, cast metal, or wire.

Most of the accessory manufacturers are small firms because highly skilled crafts are involved in production, however, the shoe and hosiery industries are dominated by large companies. Domestic production of shoes, handbags, gloves, and scarves has been overshadowed by imports from countries where labor costs much less.

The fur industry produces coats, jackets, and hats as well as trimmings for textile and leather apparel. The largest centers of fur production are Hong Kong, Canada, and Greece. There are two methods of fur production: skin-on-skin and letting-out. Because the latter is especially tedious and time consuming, the coats and other fur garments so produced are very expensive.

CHAPTER REVIEW

Terms and Concepts

Briefly identify and discuss the following terms and concepts:

1. last
2. line builder
3. modelista
4. MICAM
5. dummy
6. Hermès
7. millinery
8. gemstones
9. cabachon stones
10. lapidary
11. vermeil
12. casting
13. Swatch
14. letting-out method

Questions for Review

1. Why has fur manufacturing remained basically free of mass-production techniques?
2. Discuss the differences between the two main fur manufacturing methods.
3. Why must an accessory manufacturer pay close attention to fashion forecasts?
4. Describe shoe design development.
5. Explain the major steps involved in shoe production.
6. How are computers used in mass shoe production?
7. Name five important shoe designers or manufacturers.
8. What is the purpose of a shoe fair?
9. Briefly describe handbag production.
10. Explain how belts are apparel driven.
11. Name three well-known hat designers.
12. What is the difference between fine millinery and inexpensive hats?
13. What is used to make the finest scarves?
14. What are the basic steps in hosiery production?
15. What materials are used to make fine jewelry?
16. Name the two types of better costume jewelry.
17. Briefly explain jewelry production.
18. What country produces 95 percent of the world's luxury watches?
19. What is the effect of imports on accessory manufacturing?

Projects for Additional Learning

1. Arrange to visit the fur salon in a department or specialty store. Ask to see and feel a variety of furs. Note the fashion styling of today's furs. How do you feel about fur as wearing apparel?
2. Visit the hat department of a local specialty store. Note the differences between hats and fine millinery and the variations between functional and fashion styling. How do the millinery looks tie in with the apparel trends in the store?
3. Look for shoe advertisements in a fashion magazine. Collect pictures of five popular dressy styles and five popular casual or sport styles. How has function affected styling?

NOTES

[1] "U. S. Industrial Outlook," U. S. Department of Commerce, January, 1993, p. 33–8.

[2] Ibid., p. 33–12.

[3] "The Changing Market Share," *Fur World*, March 15, 1993, p. 4.

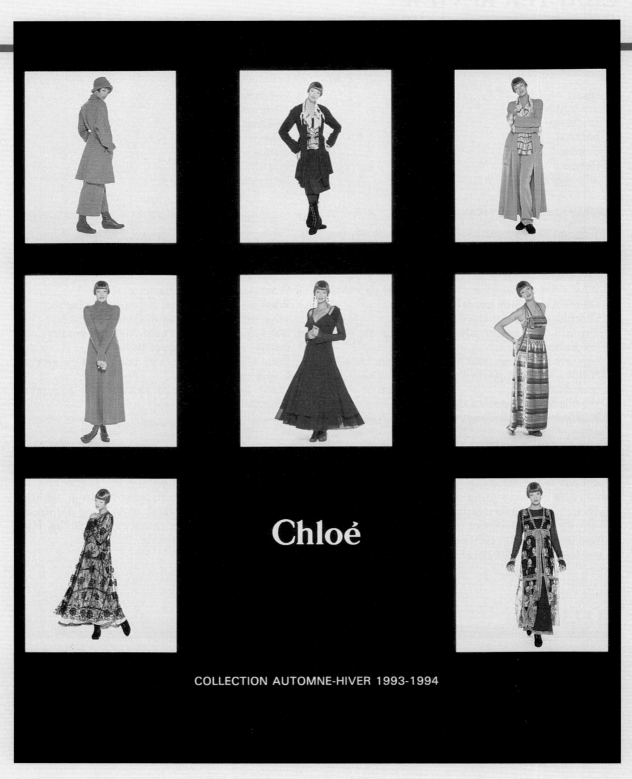

Chloé

COLLECTION AUTOMNE-HIVER 1993-1994

Chloé collection promotional photographs. *(Courtesy of the designer and photographer, Karl Lagerfeld)*

12

THE GLOBAL MARKETPLACE

CAREER FOCUS

Marketers, all those involved in the marketing process, must understand the product, the industry, the economy, world trade, and the needs of retail stores. Providing the key link to the retailer, marketing specialists try to establish contacts and a good rapport with the stores they supply.

Marketers include manufacturing management, designers, merchandisers, sales representatives, customer service representatives, merchandising representatives, all of the various people involved in sales promotion and advertising, catalog directors, publicists, writers, photographers, video producers, models, fashion show coordinators, television fashion specialists, and trade association personnel.

CHAPTER OBJECTIVES

After reading this chapter you should have attained competence in the following areas:
1. Knowledge of major international markets of apparel.
2. Understanding of the aspects of marketing.
3. The ability to explain the impact of imports on the industry.
4. Knowledge of international trade regulations.
5. The ability to comprehend collection openings, line releases, and market weeks.
6. Understanding of distribution policies.
7. The ability to describe various aids to selling.
8. The ability to discuss the forms of sales promotion.
9. Understanding of the contribution of computer technology to the field of marketing and distribution.

*I*nternational wholesale apparel and accessory markets are the means of distributing the manufacturer's finished products to retailers. A market can be any of several things: a potential demand, a place, an area, or even a period of time.

A *market* is the potential demand for a product, or the place, area, or time at which buyers and sellers meet to transact business. In Chapters 2 and 4 we discussed consumer demand and market research. Manufacturers must know whether there is a market for their products. Conversely, they must develop products to answer demand and establish and maintain a truly unique market niche.

In this chapter we refer specifically to the wholesale fashion market, where the sellers represent fashion manufacturers and the buyers are fashion retailers. Traditionally, fashion markets have been located close to suppliers and manufacturers; therefore, most market centers are also production centers. However, newer market centers have been created in various locales for the convenience of buyers.

We refer to the domestic market (the United States), to the regional markets within it, and to the international or global market. This chapter also discusses international fashion centers and their role in buying and selling on the wholesale level.

Although it is usually possible to buy year round, specific weeks are scheduled for collection openings or for the purpose of bringing buyers and sellers together. For this reason, a market calendar is included in this chapter.

Fashion marketing involves the entire process of research, planning, promotion, selling, buying, and distribution of goods. An understanding of marketing will help us to understand the interrelationship and the interdependence of various levels of the fashion industry.

GLOBAL TRADE

The fashion business has become a global phenomenon, and it is important for everyone in fashion marketing to understand all aspects of international trade.

The internationalization of fashion has created a global point of view and influenced the marketing of all products and services. Manufacturers such as Levi Strauss, Benetton, and Escada have adopted global marketing strategies. American designer Bill Blass explains, "we must be worldwide contenders in this fashion game."[1]

World trade in apparel and accessories is growing despite high tariffs, an elaborate system of quotas, and drastic currency fluctuations. Many countries may be involved in the production of a single garment. For example, a garment could be designed in France of Italian fabric, made in Hong Kong, and then distributed all over the world by importers who make the goods available to retail store buyers.

Imports

Imports are goods that are brought in from a foreign country to sell. The people or firms that import goods are called *importers*. Manufacturers and retailers can also be importers. There are two types of imports:

Imported fashion merchandise. The first type is the importation of fashion merchandise designed and produced by foreign designers and manufacturers and purchased by retailers at international markets. The United States has long imported fashion from Paris, woolens from the British Isles, sweaters from Scandinavia, and leather goods from Italy. Currently, the European Economic Community, followed by the United States, is the largest importer and consumer of apparel in the world.

Imported production. The second type of import occurs when manufacturers become importers and contract production in a foreign country where labor is cheaper. More than 61 percent of the apparel purchased in the U.S. is made either entirely abroad or made of imported fabric.[2] This huge increase in imports in recent years has stimulated a great deal of controversy. The controversy revolves around two key points: the balance of trade and the loss of jobs at home.

Balance of trade

The *balance of trade* is the difference in value between a country's exports and its imports. Ideally, the two figures should be about equal. Lately, however, the United States has been importing much more than it exports, sending American dollars abroad to pay for these goods and creating a huge *trade deficit*. To try to offset this imbalance, *duties* (tariffs) are levied on imports. However, even with duties, many imports can still be purchased for less than the same merchandise produced domestically. Many people feel that we should import less goods to balance trade.

Others feel that the consumer should be given the best merchandise at the best price regardless of the balance of trade.

Labor versus free trade

Another controversy between labor and proponents of free trade concerns the second type of importing: offshore or overseas production. There has been an enormous increase of imported apparel and accessories especially from China (which has replaced Hong Kong as the largest foreign apparel supplier to the United States), the Caribbean Basin, India, Indonesia, Malaysia, Thailand, Sri Lanka, Bangladesh, and Macao.[3] Furthermore, manufacturers tend to use fabrics from the region where production is done, thereby taking sales away from their own textile producers. Labor unions complain that overseas production steals thousands of domestic jobs in textile and apparel production. Retailers argue that workers overseas produce quality merchandise at lower prices, offering American consumers a wider choice. Because both arguments have validity, a solution to the problem is difficult to resolve.

Many marketing experts and proponents of *free trade* (trade without restrictions) believe that in the long run it would be best if world trade were based on specialization. That is, each nation would contribute to the world market what it produces best at the most reasonable cost. In this way consumers would obtain the most value for their money as well as a wide choice of merchandise from around the world. Designer Elie Tahari has said, "There should be free trade all over the world. Protectionism only limits growth and creativity."[4]

Import tariffs or duties

Duties or tariffs are customs charges imposed on imports in an attempt to protect domestic industry. Due to much lower wages, it costs less to produce garments or accessories in undeveloped countries than in the United States, Canada, or Western Europe. The duty varies according to the type of garment, its fiber content, and whether the fabric is woven or knitted.

Harmonized Tariff Schedule (HTS)

The Harmonized Tarriff Schedule is an international system of product classification used since 1988 for international customs clearance and the collection of data on imports and exports. Formerly, every country had its own system of naming or numbering imports and exports. Now a system of 6-digit codes allows participating countries to classify traded goods on a common basis.

Quota allocations

Many governments also regulate imports by means of *quotas*,regulations which control the quantity of imported merchandise. Quotas are negotiated agreements between two trading countries and are allocated for a 12-month period. Before considering overseas sourcing, a manufacturer or importer must be sure that there is enough quota to allow the merchandise in question to be produced in that country. Frequently, all the allocations are already held by manufacturers or by governments in a given country. "Capturing quota" is often a determining factor as to where goods are made.

The General Agreement on Tariffs and Trade (GATT)

Created in the United Nations after World War II, the purpose of GATT was to stimulate economic growth in war-torn and developing countries and to reform the global trading system that was plagued by high tariffs and import quotas. The agreement is a contract among 103 governments which together account for about 90 percent of world merchandise trade.[5] Its basic objectives are to achieve the expansion of trade, the reduction of barriers to such trade, and the progressive liberalization of world trade in textile products. Some of the functions of GATT are to set rules governing trade behavior, resolve disputes between members, and serve as a forum for trade negotiations.

Members of GATT have met at eight negotiating conferences known as *rounds* to establish policies to encourage free trade and to abolish tariffs and quotas. As member countries have undertaken more complex issues, and negotiations have become more complex, rounds can last several years. At the Uruguay Round, begun in 1986, issues included liberalization of foreign investment and the protection of intellectual property such as computer software, trademarks, and videos.

Multifiber Arrangement (MFA)

Under the jurisdiction of GATT's textile committee, the MFA was negotiated in 1974 as a transitional arrangement to bridge the gap between previous trade restraints and the trade liberalization policies of GATT. It was intended only as a temporary measure to protect U.S. apparel and textile industries from import surges, but it has been renewed five times. It is proposed to phase it out over a ten-to-fifteen year period.

Exports

To help balance imports, there is a growing trend for American manufacturers to sell their merchandise to foreign retailers either through direct export or by licensing arrangements. Import and export duties and restrictions have kept this business small, but when the U.S. dollar loses value relative to other currencies, exporting becomes especially profitable. In some cases the government offers incentives, such as tax breaks, tax deferral, and low-interest credit rates, to make exporting even more attractive.

The ending of trade barriers within the European Community (EC) in 1992 has created marketing opportunities for the United States. The promise of 380 million more consumers (750 million including the Eastern countries) has made U.S. manufacturers eager to export to Europe. An increasing number of American manufacturers are exhibiting at industry trade shows in France, Switzerland, Germany, Italy, the United Kingdom, and Japan.

American designer boutiques in Europe and events such as Expo Viva America at Au Printemps, a Paris department store, promote awareness of American fashion. Some of the manufacturers already exporting to Europe are Liz Claiborne, Donna Karan, and Levi Strauss. However, American designer fashion is a hard sell to Europeans who recognize only a few

American designer or brand names. A Made in the USA label carries more weight in Asia and Latin America. So far the United States has had the most success exporting to Canada, Mexico, and Japan.

North American Free Trade Agreement (NAFTA)

If ratified, NAFTA will provide for free trade between Canada, Mexico, and the United States, creating a free market of approximately 400 million people in North America. The agreement hopes to promote economic growth through the expansion of trade and investment opportunities within the free-trade area.

Canada and Mexico are already the largest trading partners of the U.S. and, in turn, the United States accounts for more than two-thirds of their total trade. Since the Canada/U.S. Free Trade Agreement went into effect in 1989, trade between the two countries has increased. However, Canadians feel they are not well served by the agreement because of its *yarn-forward* clause which requires apparel to be made with domestic yarns and fabrics. Canadians have built a reputation around styles made with European fabrics.

Mexico stands to profit enormously from NAFTA. Mexico is already a major sourcing spot for the United States; under NAFTA it could become even more important to the U.S. than the Far East. Importers could monitor production quality and receive shipments within days rather than weeks. In addition, Mexico stands to become a greater consumer country as their standard of living improves. To protect sourcing that has been developed since the Caribbean Basin Initiative (CBI) began in 1987, it is hoped that Caribbean countries will be given the same preferred treatment as Mexico.

Opponents of NAFTA say that it would be devastating to our economy. Proponents anticipate that NAFTA will enable the North American market to compete effectively with the European Community, which since 1992 has become the largest market in the world.

INTERNATIONAL MARKETS

Now that we have covered the ground rules of international buying and selling, we are ready to discuss markets, the time and place where buying and selling occurs.

After each manufacturer's collection or line is designed, it must be presented to retail buyers so they can buy for their stores. This presentation can be accomplished by various means including fashion shows, market weeks, and contacts with sales representatives.

The terms *collection opening* and *line release* are synonymous; both signal the first opportunity for buyers to see fashion merchandise for a new season. The term line release is used by most fashion manufacturers to indicate that their new seasonal merchandise is ready to be sold. Collection openings are held by designer houses and involve gala showings.

European Collection Openings

Couture

Paris continues as the world center for couture. Even Italian designers Valentino and Versace are showing their couture collections in Paris. The spring collections are shown in January and the fall collections in July. Collections are now largely shown at a new fashion center beneath the Place du Carrousel near the Louvre.

The couture is highly regulated by the French Ministry of Industry and the Fédération Française de la Couture which have recently eased the regulations governing couture. Designers now have to show a minimum of 50 pieces in a collection; new couturiers have to show only 25 pieces for a two-year transition period. *Houses* (couture companies) may now show or send videotapes of the collections to private clients instead of having 45 informal fashion shows per season, which was a waste of both time and money. Still, it costs approximately $1 to $2 million a year for a large house to produce their two couture collections because costs include fabrics, labor, especially made accessories, models, catwalk, and sound systems.

To compensate for the losses, the French government gives support to the couture in order to promote exports. Government-owned French television gives the couture free exposure. This kind of publicity is more prestigious than costly advertising and helps to generate sales of the couturiers' ready-to-wear, perfume, and licensing businesses, thereby helping to keep the couture alive.

Prêt-à-porter

European ready-to-wear openings are held twice a year: fall collections are shown in March and spring collections in September or October. They are shown earlier than couture to allow time for mass production.

Paris is such a draw that designers from other countries such as Romeo Gigli from Italy, Jil Sander from Germany, Kansai Yamamoto and Issey Miyake from Japan, and Katharine Hamnett from London show their collections in Paris. The Paris prêt-à-porter shows are generally preceded by shows in Milan. The French, Italians, and Germans want to consolidate their show schedules to make them run sequentially, possibly squeezing out London. A buyer would over two weeks to see the shows in Paris, Milan, London, Düsseldorf, and Barcelona.

Many designers already show large portions of their collections to buyers one or two months

Salon International du Prêt-à-Porter Féminin.
(Courtesy of PROFEM, Studio Photos Salon)

before the runway show. Buyers of better merchandise and approximately 2000 journalists from all over the world flock to see the individual showings of designers such as Karl Lagerfeld. These fashion shows are by invitation only and have become very theatrical. Most buyers see ten shows a day, from early morning until late at night, taking notes to remember styles. After analyzing the collections from the point of view of their stores' needs, buyers place orders, usually through *commissionaires* (representatives) or their own buying offices in each city. Deliveries are staggered so that new merchandise arrives periodically in the stores to capture customer interest.

The French prêt-à-porter shows are not limited to the designer collections. There are other shows and groups of exhibitors spread out all over Paris, organized by various fashion associations. While the Chambre Syndicale designer shows are usually at the Louvre, there is a large show at the Porte de Versailles. There are also other exhibit groups called Paris sur Mode, Atmosphere d'Hiver, and Tranoi. Some manufacturers show in hotels or restaurants. In Milan, both the Camera Nazionale della Moda Italiana designer collections and the Milanovendemoda are together at the fairgrounds with some overflow in hotels. In Germany, the first ready-to-wear show of the season is Collections Premieren Düsseldorf (CPD).

Market weeks and fairs

The collection openings are held in their respective fashion centers as part of market weeks or fairs. Markets are held in specially built market buildings called "fairgrounds" or in hotels. The market can last from three days to two weeks. Many Americans do not understand that European trade fairs are markets and not just expositions. In Germany, businesses often sell 75 percent of their merchandise at these markets. Markets are an excellent opportunity for the manufacturer to reach new stores, establish new accounts, and, in turn, help retailers looking for new resources.

The German fashion fair IGEDO (Interessensgemeinschaft für Damenoberbekleidung—interest group for women's wear), in Düsseldorf, which also produces the CPD shows, has developed into the world's largest women's wear market. Manufacturers from 73 countries show their fashion lines to over 200,000 buyers each year at four markets.

Markets of men's clothing are held in Paris, Florence, Milan, London, Cologne, and Copenhagen in January or February and again in August or September. There are collection openings, fashion fairs, or market weeks going on somewhere the year round. Table 12–1 presents a calendar of the world's most important fabric, apparel, and accessories fairs and market weeks.

DOMESTIC MARKETS

American Collection Showings and Line Releases

American collection showings and line releases for women's better fashion usually occur five times a year:

◆ Summer merchandise is shown in January

◆ Early fall merchandise is shown in late February or early March

◆ Fall II merchandise is shown in late March or early April
◆ Holiday and/or resort merchandise is shown in August
◆ Spring merchandise is shown in November

These shows are followed by regional markets. However, most lines stagger delivery dates of various groups so that new merchandise frequently comes into the store. Children's wear, formerly shown only twice a year, has now become so fashion-oriented that it is also shown five times a year, often at the same time as women's wear.

Men's wear is traditionally shown twice a year, for fall and spring. However, releases are becoming less structured as men's fashion becomes more style oriented. The Clothing Manufacturers Association holds its market week for tailored suits and coats in late January or February for fall and in late August or early September for spring. The National Association of Men's Sportswear shows fall lines in March or early April and spring lines in October. These shows are followed by regional markets.

New York

New York remains the major fashion market of the United States. Seasonal designer collections are usually presented as fashion shows. The Council of Fashion Designers of America (CFDA) is organizing members to show together in large tents in Bryant Park adjacent to the New York Public Library. Out-of-town buyers and journalists attend shows and exhibits which are usually mounted in the designer's showroom, in hotels, or in rented spaces around New York City. Buyers from large department and specialty stores travel to New York because many popular lines are "sold up" before any regional markets begin. Liz Claiborne, for example, sells all of their sportswear out of the company's New York showroom.

An Oscar de la Renta collection showing. (*Courtesy of Oscar de la Renta*)

TABLE 12-1
Important International Fashion Markets
and Collection Showings
(Some of the hundreds of fashion related markets happening year round throughout the world)

January

Pitti Uomo and Uomo Italia (men)	Florence
Milano Collezioni Uomo (men)	Milan
Designer Men's Wear Collections	Paris
Fashion Accessories Expo	New York City
Tokyo Fashion Week	Tokyo
Hong Kong Fashion Week	Hong Kong
NAMSB Show (men's)	New York
Couture collections (spring)	Paris
New York designer market (women's r-t-w for summer/transitional)	New York City
Los Angeles R-T-W market (summer)	Los Angeles
Salon International du Prêt-à-Porter Féminin (into February)	Paris

February

Salon de la Mode Enfantine (children)	Paris
SEHM—Salon International de l'Habillement Masculin	Paris
Herren Mode Woche (men)	Cologne
IMBEX—International Men's and Boy's Wear	London
CPD— Collections Premieren Düsseldorf	Düsseldorf
FFANY—Fashion Footwear Association of New York and National Shoe Fair	New York City
Premier Collections and The London Show (women's r-t-w)	London
Imagenmoda (women's r-t-w)	Madrid
Mode Woche München (women's r-t-w)	Munich
FACE—Fashion Apparel Collections Exhibition (women's r-t-w)	Montreal
Canadian Designer Collections (women's r-t-w)	Toronto
New York designer market (women's r-t-w for fall)	New York City
Paris Designer Collections (women's r-t-w) (or March)	Paris

March

Première Classe (accessories)	Paris
Milan Designer Collections and Milano Collezioni Donna (women's r-t-w)	Milan
IGEDO (women's r-t-w)	Düsseldorf
Premier Vision (fabrics)	Paris
MIDEC—Mode Internationale de la Chaussure (shoes)	Paris
MIPEL (leather)	Milan
GDS (shoes)	Düsseldorf
MICAM (shoes)	Bologna
SIIF—Fur Industries Salon	Paris
Fur and Fashion Frankfurt	Frankfurt
New York designer market (women's r-t-w for fall II, holiday) (into April)	New York City
Ideacomo (fabrics)	Como

April

American Designer Collection Shows (women's r-t-w for fall)	New York City
Los Angeles R-T-W market (fall I)	Los Angeles
Interstoff (fabrics)	Frankfurt

May

Fashion Accessories Exposition	New York

June

ESMA-Eurotricot (knitwear)	Milan
Milano Collezioni (men)	Milan
Pitti Uomo and Uomo Italia (men)	Florence
Pitti Bimbo and Moda Bimbo (children)	Florence
NAMSB Show (men)	New York
Los Angeles R-T-W market (fall II, holiday)	Los Angeles

July

Designer Men's Wear Collections	Paris
Couture collections (fall-winter)	Paris
Pitti Bimbo (children)	Florence
Hong Kong Fashion Week	Hong Kong
Tokyo Fashion Week	Tokyo

August

New York designer market (women's r-t-w for resort)	New York City
Fashion Accessories Expo	New York City
Los Angeles R-T-W market (resort)	Los Angeles
CPD—Collections Premieren Düsseldorf (women's r-t-w)	Düsseldorf
Herren Mode Woche (men) and Interjeans	Cologne
Mode Woche München	Munich
FFANY—Fashion Footwear Association of New York and National Shoe Fair	New York City
FACE—Fashion Apparel Collections Exhibition (women's r-t-w)	Toronto

September

Salon du Prêt-à-Porter Féminin (spring women's r-t-w)	Paris
Première Classe (accessories)	Paris
IGEDO (women's r-t-w)	Düsseldorf
SEHM—Salon International de l'Habillement Masculin (men)	Paris
IMBEX—International Men's and Boy's Wear	London
Salon de la Mode Enfantine (children)	Paris
Semaine de Cuir (leather goods)	Paris
GDS (shoes)	Düsseldorf
MIDEC—Mode Internationale de la Chaussure (shoes)	Paris
MICAM (shoes)	Milan
IMB—International Clothing Machine Fair	Cologne
Imagenmoda (women's r-t-w)	Madrid
Paris Designer Collections (women's r-t-w)	Paris

October

The London Show (women's r-t-w)	London
Milano Collezioni Donna (women's r-t-w)	Milan
Bobbin Show (technology)	Atlanta
Ideacomo	Como
Premier Vision (fabrics)	Paris
Interstoff (fabrics)	Frankfurt
New York designer market (women's r-t-w for spring) (into November)	New York City
Fashion Accessories Expo	New York

November

American Designer Collection Shows (women's r-t-w for spring)	New York City
Los Angeles R-T-W market (spring)	Los Angeles

December

ESMA-Eurotricot (knitwear)	Milan
FFANY—Fashion Footwear Association of New York and National Shoe Fair	New York City

Sample merchandise is shown to prospective buyers in the showrooms of the Seventh Avenue manufacturers, often next to design studios. Apparel showrooms are usually grouped in buildings according to merchandise classification and price range. Grouping by apparel type was organized for the convenience of the buyer who does not have the time to travel all over town yet must see everything offered in a particular category and price range. For example, the prestigious addresses of 530 and 550 Seventh Avenue have traditionally been designer addresses. Broadway houses showrooms for moderately priced women's wear. Lingerie and intimate apparel are centered on Madison Avenue. Children's wear showrooms are grouped around Thirty-fourth Street and Sixth Avenue and at Abraham & Straus plaza at Thirty-third Street and Sixth. One large building at 1290 Avenue of the Americas has showrooms representing approximately 75 percent of domestically produced men's clothing. The Empire State Building houses showrooms of men's furnishings. The buildings are becoming less specialized, however, as showrooms relocate because of lease losses or the need for more space.

Regional Market Centers

Regional markets are used primarily by small store owners who have neither the time nor the money to travel to New York. Some of these regional markets have developed because the area is also a manufacturing center. Large buildings called *marts* have been built in these centers to house showrooms. Within the marts, showrooms are grouped according to category for the buyers' convenience. For example, men's furnishings showrooms may be grouped on one floor, women's lingerie on another, and so forth. Regional markets are known for moderate and budget apparel because of the needs of the stores that buy there. The major regional centers are:

Los Angeles. Los Angeles has become the nation's second largest fashion market center and is a showcase for California designers. The California Mart's first building opened in 1964. The recently renovated 13-story, 5-building complex now houses 2000 permanent showrooms.

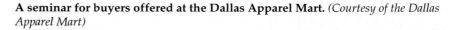

A seminar for buyers offered at the Dallas Apparel Mart. (*Courtesy of the Dallas Apparel Mart*)

Dallas. The Dallas Apparel Mart is part of the Dallas Market Center Complex. Opened in 1964, the mart now has 1200 women's and children's showrooms and 300 men's wear showrooms. This mart is expected to become a NAFTA trade center and already offers seminars on subjects such as customs procedures and international banking.

Chicago. The Chicago Merchandise Mart, serving the Midwest, has approximately 800 permanent showrooms exhibiting 4000 lines.

Atlanta. The Atlanta Apparel Mart, which serves the Southeast, now houses 2000 permanent showrooms.

San Francisco. The Fashion Center, serving the northwest, is a 6-story, atrium-style building of 740,000 square feet. The center sponsors the Golden Shears Awards to bay area designers.

Other regional marts are the Denver Merchandise Mart, Colorado; the Carolina Trade Mart, Charlotte, North Carolina; the Miami Merchandise Mart, Miami, Florida; the Northeast Trade Center, Woburn, Massachusetts; the Radisson Center, Minneapolis, Minnesota; and the Trade Center, Kansas City, Missouri.

All the regional fashion marts offer year-round market weeks, fashion shows, and educational seminars on subjects such as visual merchandising, management, and fashion show production. Other facilities include restaurants, auditoriums, hotels, hair salons, health clubs, printing services, and parking. To attract retailers and stimulate business, regional marts have developed aggressive new strategies. *Megamarts* (combined women's, men's, children's, and accessory markets) have been created for *crossover buying* so that buyers can make fewer or shorter trips to market and save travel costs. Other incentives include free or discounted hotel accommodations and air fares and more events throughout the year aimed at specific market niches.

The Showroom

A *showroom* is a place where manufacturers' sales representatives show samples to prospective buyers. In France it is called a *salon des présentations*. In high-fashion salons, fashions are modeled; in medium-to-lower-priced apparel showrooms, clothes are displayed on hangers. The showroom is outfitted with display racks, sometimes mounted on the walls for greater visibility, and with tables and chairs for the clients' comfort. Showrooms provide continual exposure for the line. They can be located on the same premises as the factory, in a mart, or they can be set up in a hotel room or market-week pavilion. The showroom is the primary setting for selling merchandise to retail buyers.

SELLING

Management and sales representatives carry out the activity of selling fashion merchandise to retail buyers.

The sales team of management and sales representatives must communicate design and merchandising concepts to retailers and may often recommend what part of the line should be purchased on the basis of the store's image,

A sales rep presenting her line to a buyer.
(Courtesy of the Dallas Apparel Mart)

customers, and needs. In turn, the sales team can give feedback to the manufacturer's design and merchandising departments regarding retail needs and line evaluation.

There are two basic ways of selling fashion merchandise to retail stores: corporate selling and selling through sales representatives.

Corporate Selling

Most major vendors, including designer companies and large moderate priced manufacturers, no longer have sales representatives. Fashion distribution to large store chains and groups has become so complex and so important to volume sales that the selling is now done *management to management.* The actual selling at this level is not as important as maintaining consistency of product and a good relationship with the store. Manufacturers and retailers are trying to build partnerships to help each other do better business. It seems the only way to survive.

Sales Representatives

Many companies still employ sales representatives ("reps") to call on specialty stores who do not have the time nor the money to travel to New York. Sales reps are salaried or paid on a commission basis ranging from 5 to 10 percent paid on orders actually shipped to and accepted by the stores. Most commissioned representatives pay all their own expenses, including part of showroom or market costs which can run as high as one-third of their income. Independent sales reps carry several small, non-competing lines. Sales representatives belong to associations that sponsor market weeks and provide other services.

After a line release, sales reps receive duplicate samples and take the line on the road to market weeks around the country and to towns and cities within their assigned geographical territories. The rep usually sets up a presentation in a mart showroom or centrally located hotel room where buyers from the area can come to see the line. To win new accounts or to introduce a new line, reps have to seek out buyers because the buyers rarely have time to look for new resources themselves.

Selling Incentives

Retailers look for manufacturers whose styling, quality, and dependability they can trust and from whom they can expect the same level of styling season after season. Retailers expect the following:

1. Consistent quality
2. Continuity of styles (a guarantee that all styles ordered will be produced)

3. On-time deliveries

4. Value

5. Reorder performance (manufacturer is able to fill reorders when needed)

However, styling, value, quality, and on-time deliveries are often not enough. Many retail executives have become more demanding with manufacturers, insisting on contributions for markdowns, advertising, and promotions. The manufacturer may offer, or the retailer may demand, one or more of the following:

◆ Incentive pricing (to allow for greater profit margins)

◆ Off-price or promotional goods (special buys at low prices, usually offered as a preselected package)

◆ Credit

◆ Markdown allowances (credit on goods that had to be marked down)

◆ Exchange or return privileges (allowing a retailer to return any unsold garments for credit or a refund)

◆ A discount of 8 percent if bills are paid on time (taken even if they pay late)

◆ Cooperative advertising allowances

◆ Promotional aids such as in-store clinics, designer trunk shows, and brochures

◆ Assistance in reordering (for those firms that reorder, reorder forms or direct reordering contact via electronic data-processing terminals may be supplied)

◆ Customer service is usually provided by manufacturers to follow up on orders and insure customer satisfaction. Liz Claiborne, for example, has 95 customer-service telephone operators who field questions from retailers.

Distribution Policy

In regards to selling, manufacturers decide on a distribution policy. Their merchandise quality and prices must be at the proper level to attract their target customers and therefore certain types of retail stores. For example, manufacturers of designer fashion sell to better department stores and fine specialty shops. Manufacturers of moderate-priced apparel and accessories sell to a wider variety of department and specialty stores. Popular and budget-line manufacturers sell to discount and other price-oriented retailers.

The manufacturer must plan distribution so that

1. The proper stores buy the merchandise,

2. The merchandise is represented in desired geographical areas,

3. One store does not create unfair competition for another, and

4. The estimated business volume is obtained.

The manufacturer may have an *open distribution* policy, selling to anyone who can pay for the goods (which is most common), or a *selected distribution* policy, limiting the number of stores in an area that may buy. When Giorgio Armani opened Armani Boutiques in West Germany, he took the Emporio Armani line away from 150 clients who had carried it before in order to keep his clothes more exclusive. Manufacturers such as Hermès and Chanel limit the actual number of garments or accessories that they produce and distribute in order to maintain exclusivity. Retailers often compete to be allowed to buy designer lines.

Brand names

Manufacturers with national distribution strive for *brand name status.* The brand name itself is very important; it must fit the image the manufacturer wants to project, reflect the style and mood of the clothes, and appeal to the intended customer. It is difficult to alter an image, as many manufacturers and retailers have discovered when they tried to upgrade their merchandise. The ultimate goal of the manufacturer is to establish the identity of a particular brand to such an extent that consumers prefer that brand compared to all others—a phenomenon sometimes referred to as *consumer franchising.* When this is achieved, brand-name recognition and the resulting consumer demand almost dictate retail buying choices. Stores that buy Liz Claiborne, for example, must support all of their concepts (buy from each group).[6] Donna Karan has minimum-order requirements that are aimed at insuring proper merchandising of the collection. Consumer franchising makes it difficult for other manufacturers in the same product area to compete because retail budgets are already allocated to the popular producers. However, while the label is an element in the consumer's decision making, it has to be backed up by quality and value.

Manufacturers support *brand integrity* with quality control, licenses kept to a minimum for purposes of control, appropriate advertising campaigns, in-store fixtures to create a consistent image, and service to retailers and consumers. Some companies such as Liz Claiborne own their own accessories divisions in order to have complete control over product and marketing.

Licensing

Licenses provide a means of diversification for the designer or brand without the risk of capital investment or the responsibility of production. Under licensing agreements, popular designers and brand-name manufacturers give other manufacturers permission to use their names and/or designs. For this the designer or brand name is usually paid a royalty of 5 to 8 percent of wholesale sales.

Licenses make it possible for designers to produce a line of coats or accessories to complement and complete their fashion statement, things they could not make in their own company for lack of expertise or capital. The licensee has expertise in production and marketing of a particular product, and in exchange gets the designer image to trade on. The licensee has experienced accessory sales representatives to work with retail accessory buyers. Anne Klein, Adrienne Vittadini, Ellen Tracy, and Nicole Miller license their names to produce accessories. There are a few designers who have hundreds of licenses: Pierre Cardin has 840, Yves St. Laurent has 200, while Karl Lagerfeld has a mere 30.[7] Ralph Lauren licenses his women's wear collection, fragrances, eyewear, hosiery, leatherwear, Chaps men's wear, luggage, and handbags. Guess licenses watches, knitwear, eyewear, infants' wear, boy's wear, legwear, and shoes.

Licensing remains a valid way to reach certain countries where trade barriers exist and to keep international manufacturing expenses low through local production. However, many designers are becoming increasingly wary of indiscriminate licensing for a broad range of products as it is impossible to control quality and image.

Joint ventures

There is a trend away from licensing towards *joint ventures* which are a form of partnership between the designer and the specialty producer. For example, a coat manufacturer and a designer could arrange a joint venture to produce that designer's coats. It is a give-and-take relationship in terms of merchandising and marketing the line. The designer has more control, yet the licensee has less risk as payment is usually on the basis of profits.

Private label

Many large retail stores are competing with manufacturers by creating their own private brands. Some stores set up their own design and merchandising staff and use contractors to produce the merchandise; others may work with manufacturers who specialize in producing private-label merchandise. They may simply copy a hot-selling item, produce basics in a wide assortment, or create an entire collection. Private labels are growing because they provide stores with the exclusivity that could draw customers from the competition (see Chapter 13 for further discussion).

Manufacturers as retailers

A *vertical company* is one that owns facilities in more than one level of the apparel or accessories industries. A vertical company may own a combination of textile plants, apparel factories, and/or retail stores. A vertical company such as Max Mara in Italy owns facilities to make fabrics and produce apparel. Vertical operations such as Brooks Brothers are fairly common in the men's wear field. There are many vertical shoe companies such as Bally or Johnston & Murphy that operate chain stores that carry their own brands. Laura Ashley produces its own fabrics, apparel, and sells them in company-owned stores worldwide. Yves St. Laurent has had his Rive Gauche shops for many years. More and more designers and popular brand manufacturers are opening retail stores that carry only their own lines. Firms such as Ralph Lauren, Anne Klein, Adrienne Vittadini, Nicole Miller, Liz Claiborne, and Esprit have opened their own stores.

By having their own stores, these firms are able to sell directly to the consumer and therefore enjoy bigger *profit margins* as there are no wholesale costs. This arrangement affords the opportunity to display a total concept, including accessories, without having it edited by a buyer, thereby setting a merchandising example for retailers. However, most retailers are not pleased that they are in direct competition with their own suppliers.

Factory outlet stores

One of the fastest growing forms of retailing is *factory outlet stores.* Formerly the only retailing done by manufacturers, outlet stores were located at production facilities where they sold *overruns* (garments not purchased by a store) or *seconds* (garments with flaws). But now most of the merchandise is flawless, deep in size and selection. Outlet stores have opened around the country in specially created outlet malls. Retailers are naturally unhappy that their own suppliers are underselling them (see Chapter 13 for the retailers' reactions).

Jobbers

In the fashion industry, merchandise is usually sold directly from manufacturers to retailers. The exception is the *jobber,* a trader who buys leftover goods from many manufacturers and resells them. These goods can be overruns or markdowns, bought at the end of a season at a large discount to clear the manufacturer's warehouse (especially if the manufacturer does not have an outlet store). They may sell in volume to small retailers or sell to the public in their own outlet stores.

Franchising

In a franchising agreement, a manufacturer sells the rights to retail a product or product line within an area. Manufacturers benefit from this arrangement because the product must be sold under the brand name and merchandised according to specifications which protect the manufacturer's image. Retailers benefit because they are guaranteed availability of stock and the right to use the brand name in advertising, supported by the manufacturer's national advertising campaign. The store has no rights to selection from the line but must carry the entire range of merchandise. Liz Claiborne franchises its collection to Harrod's in London. In many cases the entire store is franchised such as some of the Hermès boutiques.

Leased departments

Manufacturers sometimes lease space in stores to sell their merchandise, although this practice is becoming rare. This arrangement requires no selling to a retail buyer. Leasing space is especially useful in retailing types of apparel and accessories that require salespeople with particular expertise. Fur and shoe manufacturers, for example, may lease departments because their sales staff must have special knowledge of their products.

Consignment stores

Manufacturers may sell their merchandise on consignment. In this case, the retailer provides only floor space and personnel but accepts no risks for the merchandise. The merchandise is lent to the store, and the store pays only when the merchandise is sold. The manufacturer must take back any unsold merchandise and try to sell it elsewhere which, of course, is very difficult. Consignment arrangements have been used by shops featuring new designers and are used by many secondhand stores.

Catalog sales

Certain manufacturers avoid retailers altogether by mailing their own catalogs directly to the public. Emanuel, a lower-priced collection from Emanuel Ungaro, offered a 12-page direct-response catalog (in conjunction with Saks Fifth Avenue) in Mirabella magazine. Many manufacturers feel that catalog sales are the wave of the future due to today's busy life styles, and traditional retailers are understandably threatened by them (see Chapter 13 for more information on catalogs as a retailing concept).

Television sales

Television has become an increasingly important method of distribution of apparel. Home Shopping Network (HSN) began in 1985 with HSN 1, HSN 2, and Home Shopping Spree. The QVC (Quality, Value and Convenience) Shopping Channel and the QVC Fashion Channel were introduced in 1986. Home Shopping Network emphasizes low prices while QVC uses a talk-show format featuring product information. Liz Claiborne is selling apparel from its First Issue stores on QVC. Arnold Scassi designs two lines for HSN. Jewelry designer Kenneth Jay Lane, who sells his jewelry on QVC, said, "it's like department store shopping...without the crowds and the waiting in lines."[8] (See Chapter 13).

SALES PROMOTION

Manufacturers use sales promotion to make their merchandise known to prospective retail buyers and to the public.

Effective sales promotion can often mean the difference between success and failure. Promotional efforts can take the form of publicity, advertising, and other aids that manufacturers make available to retailers who buy their products.

New Collection Openings

Shows and market weeks

Press weeks are held in conjunction with collection or market openings. The extensive arrangements for these events are handled by a press attaché, a public relations consultant, or a trade association. For major collections, invitations are sent to journalists the world over, as well as to the designer or manufacturer's best customers. A *dossier* or press kit must be prepared for each guest. The kit includes a numbered list of models to be shown, photographs and sketches of a few pieces from the collection, a press release or analysis of the collection, and perhaps a biography of the designer.

Major fashion shows are very costly ventures. It costs Anne Klein approximately $150,000 to present a seasonal collection to buyers and the press. Typical expenses include 22 models for a full day at $90,000; lighting, set design, and videotaping at $22,000; accessories at $10,000; rental of the facility at $10,000; hairstylist and two assistants at $6,000; music at $2,500; two makeup artists and two assistants at $6,000; and invitations, postage, and programs at $4,000![9]

With so many lines for buyers to remember, manufacturers often use gimmicks such as distinctive invitations or souvenirs to draw attention to their products. Promotional items such as shopping bags or T-shirts with the manufacturer's name on them can serve as walking advertisements.

Publicity

Publicity is nonpaid messages made by a company to the public regarding its products, policies, personnel, activities, or services. Fashion manufacturers use in-house public relations or publicity staffs, consultants, or agencies to create publicity material and obtain *editorial coverage* of their collections in newspapers, magazines, and on television. Fashion editors of publications and television choose the material for their articles from information and photographs sent to them by the manufacturers.

Fashion programming on television is also an opportunity for designers to get publicity. "Style With Elsa Klensch" on CNN, Jeanne Becker on VH-1's "FT-Fashion Television" from Canada, and MTV's "House of Style" with Cindy Crawford offer various approaches to presenting fashion to the consumer.

Advertisers are putting more pressure on publications for editorial coverage. "All things being equal, if we are looking at two items that are equally available nationally, if one of them is an advertiser, we'd use it," says Kimberly Bonnell, Glamour's senior fashion editor. Designer Michael Kors explains why editorial coverage is so important, "Advertising establishes your image, but editorial is *it* for prestige—and sales."[10]

Advertising

Advertising is the planning, writing, producing, and scheduling of paid announcements designed to attract potential customers' attention to the manufacturer's merchandise. Manufacturers have in-house staffs or hire outside agencies or consultants to develop and produce advertising campaigns.

Trade-directed advertising

Many manufacturers advertise in trade publications such as *Women's Wear Daily* or *Daily News Record* because they wish to reach the retail buyers who read those publications.

Consumer-directed advertising

Others may advertise directly to the consumer in publications such as *Vogue, Elle, Glamour, Lears, Vanity Fair* or the *New York Times Magazine*. Some manufacturers, such as Liz Claiborne, also advertise regionally in publications such as the *Texas Monthly*, the *Chicago Magazine* or the *Los Angeles Times*.

Brand-name advertising

Large manufacturers of apparel, hosiery, and shoes advertise their brand names nationally or globally. Kenar Enterprises is advertising in Italy even though they don't sell there! "We want to create a demand so that when we show up, they'll know who we are," explains Kenneth Zimmerman, president.[11] Brands use both trade and consumer media, thereby reaching both markets. Television has become increasingly important as an advertising medium for the manufacturer. Television ads are repeated over and over and trade newspaper ads cover many pages in order to achieve *brand-name saturation.*

Designers and manufacturers may use *image advertising* to sell their names or brands. Image advertising tries to capture the spirit of the product. Fashion changes so quickly that it is often impossible to produce an ad for a specific style for national media. In addition, since each store carries different styles from any one line, one specific style cannot be advertised for all stores.

Due to the recession, there has been a trend toward *item advertising.* Designers choose one outstanding style from their collection to picture in an ad. With image advertisements, they are able to see direct sales response. Dana Buchman, for example, advertises an item twice a year in consumer magazines such as *Vogue, Elle,* and *Harper's Bazaar.*[12]

Fashion photographers such as Irving Penn, Steven Meisel, and Bruce Weber, and popular models including Claudia Schiffer, Cindy Crawford, and Kate Moss, have commanded huge sums of money to help shape company images. Karl Lagerfeld commented that, "Each one with his [or her] own personality helped to shape the visual image of today's fashion."[13]

Cooperative advertising

Many manufacturers cooperate financially with textile producers and retailers on advertisements in order to make the public more aware of brand names. The manufacturers offer money to retailers who advertise their styles. Manufacturers' ads may also be funded by textile producers if their fibers or fabrics are used. Textile and apparel manufacturers' co-op allocations may provide a retailer with up to 50 percent of its media costs. Co-op allowances are based on the percentage of net sales to the retailer. The cooperative ads must carry the names and *logos* (brand or store symbols) of each contributing company.

Other Promotional Aids to Retailers

Manufacturers often provide retailers with aids they can use in their advertising, publicity, and public relations. Each manufacturer constantly tries to develop more selling and promotional tools. A manufacturer may offer one or more of the following to stores that purchase its merchandise:

Personal appearances.
Many designers make personal appearances at retail stores at the beginning of the major seasons to draw crowds. The appearance may be accompanied by a fashion show, talk, and/or luncheon.

A cooperative advertisement between Jessica McClintock, the manufacturer, and Ups & Downs, the retailer. *(Courtesy of Jessica McClintock)*

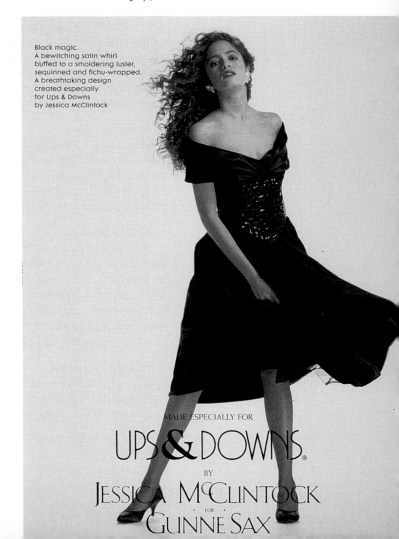

Black magic.
A bewitching satin whirl
buffed to a smoldering luster,
sequinned and fichu-wrapped.
A breathtaking design
created especially
for Ups & Downs
by Jessica McClintock

MADE ESPECIALLY FOR
UPS & DOWNS.
BY
JESSICA McCLINTOCK
FOR
GUNNE SAX

Designer trunk shows.

This is a similar idea to the personal appearance but in this case the designer brings the entire collection. Trunk shows are hard work but very successful because the line is not edited by a buyer. There is direct payback because customers can order garments in their own size immediately after a fashion show. Small manufacturers who cannot afford national advertising have found this method to be very profitable for them. There are even some couturiers trying this and arranging for fittings to be done in the stores. Philippe Venet travels to New York regularly where he shows his designs to private customers by appointment and then returns to Paris to make the clothes in his own ateliers.

In-store clinics.

Many manufacturers have discovered that their merchandise sells better if it is thoroughly explained to both salespeople and store customers. Therefore, the designer or a manufacturer's representative may visit stores to train and educate the sales associates and/or the customers with demonstrations, slides, and a talk.

Merchandise representatives.

Because manufacturers feel that stores have failed in their responsibility to train sales associates properly, some have hired consultants to work in the retail store. Some work exclusively for a large store while others are regional merchandisers servicing a cluster of stores in a particular area. Anne Klein calls them "retail sales executives," Ellen Tracy's are "retail merchandisers," and Liz Claiborne calls them "selling specialists." They are trained by the manufacturer to educate both sales associates and customers on how the line should be merchandised, worn, and accessorized. They also check to see that merchandise is displayed properly. They are in the position to give feedback to the manufacturer on competitive lines and customer reactions. It is very costly to maintain these consultants and some manufacturers have cut back on this program.

Videos.

Another trend in fashion promotion and training is to show the collection on video. Specific videos for training purposes demonstrate selling techniques and how pieces work together. Couturiers send videos to their best customers. Accessory firms use videos to show customers in a retail store how to drape a scarf or wear a hat.

Image books.

Another way to communicate an image, these booklets are mini catalogs that try to give a visual impression of a collection. They may be distributed to retailers, the press, and to customers. Dana Buchman and Adrienne Vittadini offer *look books* for customers and/or sales associates to learn about the product and how pieces work together. Depending on the size and number of books sent out, which can be anywhere from 5000 to 250,000 per season, image books cost between $40,000 and $100,000 to produce.

Display fixtures.

Manufacturers sometimes provide stores with fixtures to enhance visual merchandising. Ellen Tracy, for example, offers stores an *enhancement package* including mannequins and signing.

In-store boutiques.

Some designers are so popular that they are able to demand that their apparel and/or accessories be displayed in a specific strategic location

Pages from an Adrienne Vittadini Look Book for customers and sales associates to learn how pieces from the collection work together.

with a prescribed décor. These designers or manufacturers may provide décor and fittings for the entire shop on a cooperative basis.

Radio scripts and TV commercials.
Stores can "tag" media spots provided by the manufacturer with their own names and run them in their local areas.

Glossy photographs.
Photographs of merchandise may be provided to stores to be used for publicity or advertisements. Couturiers also send photographs, sketches, and swatches to their best customers.

Statement enclosures.
Mailing pieces can be provided for stores to send to customers in their monthly mailings.

Hang tags.
To help carry out a designer or brand image, a manufacturer's advertising department or agency may create consistent packaging including hang tags for their garments or accessories.

Associations that Promote Fashion

Trade associations support manufacturers in promoting particular segments of the industry. The American Apparel Manufacturers Association (AAMA), the Council of Fashion Designers of America (CFDA), the Fashion Footwear Association of New York (FFANY), and the Men's Fashion Association of America, are examples of the specialized organizations that exist.

The International Fashion Group, Inc., founded in 1931, fosters the careers of women in the industry. Primarily promotional and educational

Karl Lagerfeld and Isaac Mizrahi at the Council of Fashion Designers of America Awards Ceremony.
(Courtesy of CFDA, photo by Mary Hilliard)

in purpose, it now has over 5000 members with regional chapters in 24 American cities and five foreign countries.

Fashion Awards

Fashion awards presented by various organizations generate interest in fashion because of the publicity they create.

The Council of Fashion Designers of America presents annual awards in a mimimum of four categories: men's, women's, accessories, and the Perry Ellis Award for new fashion talent. The winners are chosen by a committee of fashion editors and retailers. The CFDA has also organized "7th on Sale" to raise money to fight AIDS. Held every other year, this event took place in New York in 1990, in San Francisco in 1992, and in Los Angeles in 1994.

The Dallas Fashion Awards, which began in 1976 as a local awareness program, has evolved into a recognition ceremony to honor American designers. Three designers are nominated in many categories and then national retailers can vote for the winners by mail. At the awards dinner, which is a charity benefit, a Fashion Excellence Award is presented to an outstanding figure in retailing.

COMPUTER-AIDED MARKETING AND DISTRIBUTION

As we have seen at every level of the industry, computers have become indispensible in marketing and distributing fashion.

Computers monitor inventory, sales, and customer service. The computer keeps an inventory of what fabric has been ordered, what cuts have been issued, what garments are in production, and what finished garments are in stock. This data lets merchandising and operations staff know which fabrics and colors are selling and which are not so that production and fabric purchases can be adjusted.

Quick Response (QR)

Quick Response uses computer-based information in an effort to speed up the process of ordering, designing, producing and shipping to the retailer. It enables the apparel industry to adapt to changes in the marketplace more quickly and therefore compete with imports. Quick Response initiatives encourage close working relationships among fabric suppliers, manufacturers (vendors), and retailers via *electronic data interchange (EDI)*, a computer-to-computer exchange of information. Electronic partnerships are so efficient that they have replaced the traditional long-term personal relationships that existed between vendor and buyer. All goods are given *universal product codes (UPC)* whch identify style, color, size, price, fabrication, and vendor. These bar codes are laser-scanned and the information is fed through the EDI pipeline.

Electronic Data Interchange can put through a purchase-order transaction from retailer to vendor. Through computer networks, orders can be entered via telephone from anywhere in the world. The computer can instantly report on what is in stock and what delivery dates a customer can expect based on the information that it has sorted and filed.

Computer technology also helps the customer service department follow up on a sale by automatically printing invoices, handling reorders, and even premarking for retailers. It keeps track of shipments and adjusts inventory records automatically. Some manufacturers such as Liz Claiborne use a round-the-clock network called System Updated Retail Feedback (SURF), which provides up-to-the-minute reports of what's selling and what isn't throughout the country.[14] This system enables the company to continuously monitor selling patterns and alter production accordingly.

At the end of a season, merchandisers and sales managers use computer sales records to compare actual sales figures with the original plan to see if sales and profit goals were met. These records also give merchandisers and designers specific information on what colors, styles, fabrications, and price ranges sold best which helps them plan future lines.

The impact of Quick Response has been closer-to-home sourcing and a trend toward developing better partnerships with vendors. Electronic Data Interchange tends to transfer inventory risks to manufacturers away from retailers. Retailers don't want to commit to buying merchandise too far in advance of the selling season and manufacturers try to accommodate them.

SUMMARY

The retailer and the manufacturer, who represent two separate businesses, communicate through wholesale markets. Buyers and the press are interested in seeing the newest seasonal offerings of many manufacturers. Fashion showings and markets are held in market centers all year long, around the world. An important trend in marketing is internationalization as demonstrated by the tremendous growth of imports, exports, and multinational fashion companies.

Manufacturers have various distribution policies to ensure that their goods are merchandised properly. Working through sales representatives, many offer incentives to retailers to buy their merchandise. They also use publicity, advertising, and other promotional aids to foster sales.

Computer technology has helped suppliers, manufacturers, and retailers form partnerships to try to speed turn-around time for production and sales and keep records of inventory and sales.

The final test for fashion merchandise is whether the consumer buys it at the retail level. Each season brings a new chance for success or failure. No wonder the apparel industry is a competitive and often nerve-racking, but exciting business.

CHAPTER REVIEW

Terms and Concepts

Briefly identify and discuss the following terms and concepts:

1. Markets
2. Imports
3. GATT
4. NAFTA
5. Line release
6. Salon de Prêt-à-Porter Féminin
7. IGEDO
8. Balance of trade
9. Showroom
10. Mart
11. Selected distribution policy
12. Corporate selling
13. Private label
14. Vertical companies
15. Franchising
16. Trunk shows
17. Cooperative advertising
18. Image advertising
19. Merchandise representatives

Questions for Review

1. Define the terms *market* and *marketing*.
2. How is the fashion market affected by imports?
3. What are the pros and cons of NAFTA?
4. What are two regulations regarding couture showings?
5. Why did regional and international fashion markets develop?
6. How does the presentation of a high-priced collection differ from that of a moderate or low-priced line?
7. What are two possible distribution policies?
8. How do manufacturers act as retailers?
9. List five promotional aids that manufacturers offer to retailers.
10. Explain how computer technology aids marketing.

Projects for Additional Learning

1. Analyze the imported fashion merchandise and accessories at a large retail store. Make a list of the countries represented and the merchandise each country specializes in.
2. Interview a local store's fashion buyer. Ask how and when he or she buys fashion merchandise: At a market? From a sales representative? Through a buying office? From a catalog? What type of merchandise is purchased in what way? What is the buyer's favorite method, and why? How often does the buyer go to a market center? Which one does he or she attend? Summarize the answers in a written report.
3. Ask a buyer to let you sit in when a sales representative is showing a line. Evaluate the merchandise in terms of the store's needs. Note the buyer's reactions to the line and whether he or she actually writes an order. Summarize your findings in a written report.
4. "Sell" the sportswear line you designed (Chapter 9) to your class, explaining its concepts and features.

NOTES

[1] As quoted by Zina Sawaya in "Avenue Montaigne It Isn't," *Forbes*, October 28, 1991, p. 130.

[2] M. L. Cates, Jr., American Textile Manufacturers Institute, "ATMI Calls 92 a Better Year," *Women's Wear Daily*, December 30, 1992, p. 8.

[3] "U.S. Industrial Outlook 1993-Apparel and Fabricated Textile Products," U.S. Department of Commerce, p. 32–7.

[4] As quoted in "Thumbs Up for NAFTA," *Women's Wear Daily*, August 19, 1992, p. 14.

[5] "GATT, What It Is, What It Does," Geneva, 1991, p. 1.

[6] Robert Abajian, Design Director, Liz Claiborne, Interview, May 1, 1992.

[7] "No More Easy Money," *Women's Wear Daily*, May 11, 1992, p. 5.

[8] As quoted by Maryellen Gordon, "TV: Fashion's Next Frontier," *Women's Wear Daily*, January 6, 1993, p. 9.

[9] Nancy Lueck, Anne Klein, Interview, March 1993.

[10] As quoted by Teri Agins, "Editorial Plugs for Apparel Are in Style," *Wall St. Journal*, October 6, 1992, sec. B, p. 1.

[11] As quoted in "Marketing/Media," *Women's Wear Daily*, June 12, 1992, p. 16.

[12] Gail Cook, President and CEO, Dana Buchman, Interview, April 28, 1992.

[13] As quoted by Lisa Lockwood, "Fashion's Top Photographers," *Women's Wear Daily*, March 27, 1992, p. 12.

[14] Jay Margolis, President, Liz Claiborne Sportswear, Interview, April 29, 1992.

Part Four

THE RETAILING
OF FASHION

The women's millinery department at I. Magnin, San Diego.
(Photo by Kim Brun Studios, courtesy of I. Magnin)

13

RETAIL STORES

CAREER FOCUS

One of the advantages of a career in retailing is that it offers so many people the opportunity to work close to home. Retail store organization begins with a chief executive officer (CEO), director of stores, individual store managers, floor and area managers, department managers, and their assistants. Retailing careers are also possible in mail order and television. Other possible retail careers will be discussed in the next two chapters.

CHAPTER OBJECTIVES

After reading this chapter you should have attained competence in the following areas:
1. Understanding of today's retail situation and trends.
2. Ability to discuss the various types of retail stores.
3. Knowledge of the organizational differences between single-unit and multiple-unit stores.
4. Identification of at least 10 major international stores and three famous shopping areas.
5. Ability to compare the organizational structure of a small store with that of a large store.
6. Comparison of direct retailing, vertical retailing, and private-label merchandising.
7. Appreciation of the growing importance of non-store retailing.
8. Awareness of how a store's fashion image is conveyed to consumer groups and manifested in store policies.

*R*etailing is the link between the manufacturer and the consumer. Retailers buy fashion merchandise from *vendors* (manufacturers) all over the world and bring it to their stores for convenient selling to consumers. There are nearly 2 million retail firms throughout the United States alone. About 135,000 of these retailers specialize in fashion apparel and accessories; another 70,000 include some apparel and accessories among their merchandise. True success in the fashion business is finally achieved at the retail level, by consumer acceptance measured in purchases.

Many factors go into the making of a successful retail store: good management, a convenient location, a pleasant atmosphere, exciting and appropriate stock, buyers with an understanding of customer needs, helpful salespeople, and customer service. The first half of this chapter examines the current retail situation, global retailing, types of fashion retail stores, nonstore retailing, and the organization of single-unit and multiple-unit stores. The second part of the chapter covers marketing and the establishment of a retailer's fashion image. Chapter 14 discusses retail merchandising (the buying and selling of merchandise) and Chapter 15 studies promotion. It is important to have read Chapters 2, 8, and 12 before reading this chapter.

THE RETAILING PICTURE

Enormous changes have occurred in retailing methods, management, and ownership which will have long-range implications for retail careers.

The Growth of Consumer Credit

In the 1970s and 1980s demographics were favorable for retailing, particularly with the influx of women into the workforce. The growth of consumer credit, formerly limited to department store charge accounts, and the proliferation of credit cards fostered the growth of retailing. As Leslie Wexner, chairman of The Limited, explained, "Consumers were energized because they not only had the desire to buy, but the capacity to spend."[1] However, the momentum of growth in the 80s resulted in overexpansion of stores facilitated by overborrowing.

Debt Causes Retail Crisis

Retailers took on debt to expand operations (through internal growth or by acquisition of other companies), only to discover that when the recession hit and sales declined, they did not have enough money to both repay their debts and operate their businesses. Retailers also incurred debt with *leveraged buyouts* (LBO's, see Appendix Two) in order to transfer control back into private hands. These buyouts were often based on highly optimistic cash-flow forecasts which did not materialize especially during the recession of the early 90s.

The debt incurred in buyouts forced store management to make buying and selling decisions based on debt reduction at the expense of long-range market share. Highly leveraged retailers found it necessary to sell off some of their acquisitions to pay off their debts, were forced into bankruptcy (and had to be restructured), or went out of business. Across the country, once-solid companies have closed their doors, been purchased by other companies, or forced to consolidate. Macy's was the third national retailer, in a three-year period, to seek protection under bankruptcy law. Many retailers are still consolidating and restructuring and do not know what the future holds.

Besides restructuring and consolidating, retailers face many other problems in the 90s. High rents make it difficult for new stores to get started and for malls to carry a wide range of products for one-stop shopping. Competition is very keen as so many stores carry the same merchandise. Consumers want value, convenience, and fair prices and often seek out discounters, outlet stores, and warehouse clubs. Jay Margolis, former vice chairman of Liz Claiborne, remarked, "The impact of discounters, such as Wal-Mart, on retailing has taught us that while excitement may come from designer merchandise, sales are at lower prices."[2]

Retail Strategies to Compete in the Nineties

Retailers are taking various approaches to compete in the nineties. Four of these directions seem to be the most successful: value, service, uniqueness, and non-store retailing.

Value-directed retailing

After the excesses of the eighties, the nineties are often referred to as the *value decade* or the *less decade*. To be successful and improve *productivity*(sales per square foot) a retailer must create a competitive advantage and set itself apart by offering something special. Bob Connolly, Senior Vice President of Merchandising at Wal-Mart, feels that, "The winning retailer in the 90s will have the proper price/value and assortment relationship. [Retailers] must give the customers what they want, where they want it, [and] when they want it, at a price they think is fair."[3] To have a consistent pricing policy, retailers are attempting to reduce expenses and become efficient in order to keep margins and, therefore, prices down. Value-oriented retailers include discounters like Wal-Mart, warehouse clubs, and outlet stores.

Service-oriented retailing

Retailers are attempting to become *consumer driven* (to anticipate and focus on the needs of their customers); in fact, to exceed customer's expectations. They are trying to make shopping more convenient and friendly so that customers will enjoy shopping. Many retailers are using capital investments to renovate existing units to create a warm, friendly atmosphere. This service includes maintaining in-depth stock. Service-oriented retailers include Nordstrom, Wal-Mart, and many mail-order retailers such as L. L. Bean and Lands' End.

Fashion-directed retailing

Fashion-oriented retailers focus their attention on their fashion image and offer a unique approach to merchandising, fashion leadership, and store personality. Fashion-directed retailers present up-to-the-minute fashion styling. Fashion retailers who try to be unique include Henri Bendel, The Limited, and various designer shops.

Non-store retailing

Busy people, with more important priorities, are spending less time shopping in stores and malls. To answer their needs, mail order, cable television, and computer shopping are offering consumers the convenience of shopping from the home. This strategy will be discussed under Types of Retail Operations later in the chapter.

Globalization

Just as globalization has affected manufacturing and marketing, many experts feel that one way to increase market share is through global retailing. European companies such as Laura Ashley of Great Britain, Benetton of Italy, and Escada of Germany have found the size of the American market attractive and therefore have opened shops here. Many French firms such as Chanel, Hermès, or St. Laurent operate boutiques in the United States. Fendi's 20,000 square foot shop on Fifth Avenue is the largest European designer-owned store in New York City.

Galaries Lafayette in Paris has branches in New York and Singapore
(Courtesy of Galaries Lafayette)

Americans are following suit. Experts predict that since Europe's economic integration in 1992, more American retailers will find that market too enticing to resist with its 325 million citizens. The Gap has opened stores in Great Britain and Germany. Several American designers have stores in Europe. There is an American Store in Moscow. Designers such as Pierre Cardin, Gianfranco Ferré, and Donna Karan are opening stores in China. Saks Fifth Avenue plans expansion in Europe and Japan. Japan has already attracted several American retailers such as Brooks Brothers, Talbots, Ralph Lauren, Barney's, and Charivari. Brooks Brothers has 31 freestanding or in-store shops in Japan.

Closer to home, Sears, Price Club, and Wal-Mart are looking to expand in Mexico. Wal-Mart is opening warehouse clubs in a joint venture with Cifra, Mexico's largest retailer.

In addition to opening stores in other countries, companies are investing in existing retailers in other countries. Marks & Spencer of the United Kingdom owns Brooks Brothers, Aquascutum in the U.K. is owned by Renown of Japan, Vroom & Dreesmann of the Netherlands has an investment in Dillard's, and so on. Most retailers feel that globalization is the only way to survive in the future.

Retailing Roots in the Cities

Most of the world's top retail stores began in manufacturing and marketing centers such as Paris, London, Tokyo, Rome, Milan, or New York. Certain city streets or areas have become famous for shopping: Fifth Avenue in New York City; the Union Square area in San Francisco; Rodeo

Drive in Beverly Hills; Faubourg St. Honoré, Avenue Montaigne, les Halles, and the St. Germain district in Paris; Via Condotti in Rome; Via della Spiga and Via Monte Napoleone in Milan; the Roppomgi and Harajuku districts in Tokyo (which claims to have more fashion stores per capita than any other city in the world) and Regent Street, Bond Street, and Knightsbridge in London. Most *flagship* (main store in a group or chain) stores have remained in the cities.

After a loss of business to the suburbs in the last fifty years, there has been renewed interest in redeveloping retail potential in U.S. cities. Many cities are conducting revitalization projects, including the refurbishing of older department and specialty stores. In Europe, cities have continued to be the centers of fashion retailing. To further encourage shoppers, many streets have been closed to automobile traffic, creating pleasant walkways between shops. As a further development, malls are coming to American cities. Among these are vertical malls such as Trump Tower and A & S Plaza in New York City, Water Tower Place in Chicago, and the San Francisco Centre.

Suburban Shopping Centers

The shopping centers which now line many of the world's highways developed as a result of the increase in suburban living which followed World War II, which was fostered by real estate developers. The shopping center was followed by the development of the mall, the closed mall, and finally the regional closed mall. The United States now has the most shopping centers, followed by Canada, Australia, England, France, Germany, Sweden, and Switzerland. Approximately 35,000 shopping centers and 1,800 larger regional malls account for nearly half of the retail sales in the United States.[4] Sally Frame Kasaks points out that "a mall is like a department store, all the specialty stores within it create the same thing."[5]

Emphasis on value pricing

The traditional mall is anchored by at least two department stores with many other small specialty stores. However, with the success of discount stores such as Wal-Mart, many new malls and floundering older malls are using Wal-Mart, K-mart, Price Club, and other discounters as their *anchors*.

Diversified malls.
A mall may be diversified with both discount and moderate specialty stores. Ted Kraus, president of TKO, a shopping center management company, explains that "There's a remerchandising of regional malls going on. You may have a mall with a Macy's wing, where you'll have upper-middle merchandise, but at the other end you'll have a Wal-Mart with lower-end stores."[6]

Value retail centers.
The second type of mall is the emerging value retail center. These malls or strip shopping centers, made up entirely of discount stores, are growing in numbers, size, and popularity. When South Hills Mall in Poughkeepsie, New York, faced overwhelming competition from the new Poughkeepsie

Galleria, they decided to convert into a discount center. Now, says U.S. Mall Manager John Mannix, "the customers who visit the Galleria walk around, they ooh and aah, they have their Slurpee at the food court, and then they come to our mall to shop."[7]

Outlet malls.

Manufacturers' outlet stores, formerly located near manufacturing facilities, are now all over the country in specially created outlet malls. Consumers travel to towns such as Flemington, New Jersey; Monterey, California; Boaz, Alabama; Rockford, Michigan; and Freeport, Maine specifically to shop at their outlet store centers. Potomac Mills mall outside Washington, D.C. has over 200 stores selling merchandise at 20 to 60 percent off regular retail prices. The Franklin Mills mall near Philadelphia claims to be the largest outlet mall with 1.85 million square feet.

Emphasis on recreational shopping

Competition has caused some malls to provide entertainment to draw customers and capitalize on the recreational aspects of shopping. After all, next to the workplace and the home, stores are the third most occupied space in America, so why not make it fun? The Mall of America in Bloomington, Minnesota includes 14 movie theaters, an indoor theme park, a roller coaster, and a miniature golf course. At Forum Shops Mall in Las Vegas, Caesars Palace Casino is the anchor usually provided by a large department store. Henry Gluck, Chairman of Caesars World, Inc. says, "Our thinking is that in the future, our best competition will be people with more bells and whistles. Just having good prices is not enough."[8]

A view of the Mall of America, Bloomington, Minnesota. *(Courtesy of Mall of America)*

TYPES OF RETAIL OPERATIONS

Many types of retail operations are continually created to try to serve customers' needs.

In order to attract customers, a store's image must appeal to the customers' life style or shopping needs. The many different kinds of retail operations to fill these needs include specialty stores, department stores, mass merchants, and non-store retailers. There are many overlaps between categories and even retail experts do not agree on how to categorize stores. Retailing is continually evolving; new categories are emerging and old ones are combining.

Specialty Stores

Specialty stores cater to a particular kind of customer by providing a narrow focus of unique merchandise for specific tastes. Specialty retailers carrying just one category of merchandise are called *single-line stores;* those carrying related categories of merchandise are called *limited-line stores.* The limited-line retailer may cater only to men, women, teens, children, the professional woman, the sports enthusiast, or large sizes—the possibilities are unlimited. Single-line stores may carry only shoes or just socks! Most specialty shops buy merchandise within a certain price range as well as in a specific category.

In Europe, the traditional small specialty shop has continually predominated in fashion retailing. Examples of international specialty stores are: Harvey Nichols, Brown's, Next, Whistles, and Joseph in London; Peek & Cloppenburg and Hettlage in Germany; and the traditional designer shops of Paris, Milan, Florence, and Rome.

Leading fashion specialty stores across North America include: Ann Taylor, Barney's, Fred Hayman of Beverly Hills; Bergdorf Goodman; The Gap; Halls of Kansas City; Holt Renfrew in Canada; I. Magnin; The Limited; Nan Duskin of Philadelphia; Neiman Marcus, based in Dallas; Saks Fifth Avenue; Talbots; and Ultimo of Chicago. There are also many designer specialty stores, both European and American, such as Hermès, Christian Dior, Yves St. Laurent's Rive Gauche, Donna Karan, and Nicole Miller. Many specialty stores

A/X Armani Exchange, an Italian specialty store, has a shop in Saks Fifth Avenue, an American specialty store. *(Courtesy of Saks Fifth Avenue)*

such as The Gap, The Limited, Ann Taylor, Gantos, Georgiou, and Fashion Bug (Charming Shoppes) have added more and more stores and have grown into chains. Ann Taylor has over 200 stores; The Gap has over 800 stores plus approximately 260 Gap Kids stores and over 160 Banana Republic stores.

Small specialty stores are having a difficult time trying to compete in the 90s. While they have emphasized personal relations with their customers in the past, today's busy consumer does not have the time to spend hours trying on clothes. Joseph Cicio, chairman and CEO of I. Magnin, remarked, "The day of the salon. . . doesn't exist anymore." He added that merchants must consider each square foot of space from a productivity perspective and not continue with the carriage trade mentality. Many observers say that the needs of today's customers aren't satisfied by the small specialty store that doesn't have the assortment that a Neiman Marcus or a Saks Fifth Avenue has, and that has never lent itself to browsing.[9]

Deep-niche retailing

There has been a trend towards *deep-niche retailing,* encouraging specialty stores to cater to a very specific consumer, in order to obtain dominance in a classification. Examples of stores with very focused offerings include the Forgotten Woman in New York with upscale fashion for large sizes; Body Options for the exercising woman; the Sock Shops; and the Tie Rack. This development is a return to an earlier period of retailing when the consumer went to individual stores for specific needs.

Direct retailing

Many designers and manufacturers have moved toward vertical retailing, which allows them to control the product from concept to final sale in company-owned stores. This system eliminates the need for wholesale sales representatives and retail buyers. This is a common arrangement in France and Italy, where designers usually have their own shops. Manufacturers like Anne Klein, Ralph Lauren, Adrienne Vittadini, Liz Claiborne, and Esprit have opened their own retail stores. Manufacturers have become retailers and retailers have become manufacturers.

Vertical integration.

Companies which own textile and/or manufacturing facilities plus their own stores are considered vertically integrated. Great Britain's Laura Ashley is completely vertical as the company prints its own fabrics. Episode, for example, is part of a vertically integrated company owned by the Fang family. Jeffrey Fang explains, "We design our product, manufacture it, put our label inside, and then retail it in our own stores."[10] Vertical men's wear operations such as Brooks Brothers and Hartmarx, and shoe firms such as Kinney are examples in the United States. These vertical retailers would also be considered private-label merchants.

Private-label merchants.

Retailers such as The Gap, Ann Taylor, and Episode are called *private-label merchants* because they produce their own clothing which is sold under the store's name. Brooks Brothers and The Gap are examples of retailers who have established the store's name as an important brand. Corresponding merchants in Europe include Marks & Spencer and Et Vous.

The Gap is a private label retailer. *(Courtesy of The Gap)*

Private-label merchants maintain their own design departments or use the services of design studios such as Mast Industries in Boston or Dominique Peclers in Paris to design their lines. These retailers usually contract production of their product or may establish joint ventures with factories.

Department Stores

The term *department store* comes from the practice of presenting many different kinds of merchandise, each in a separate section of the store. Apparel and accessories for men, women, and children are sold along with household goods such as furniture, lamps, linens, cookware, and televisions. The government defines a department store as one that employs at least 25 people and sells three categories of general merchandise: apparel and accessories, home furnishings, and household linens. The traditional definition changed when most stores dropped appliances, toys, and other products that are now handled by specialty stores. Department stores usually concentrate 70 to 80 percent of their merchandise in the moderate to upper-moderate price ranges.

Internationally, Seibu in Japan claims to be the biggest department store in the world. GUM in Moscow is the largest in Eastern Europe; Harrod's of London is the biggest in Europe, and Macy's the largest in the United States. Other leading international department stores include: Liberty and Selfridges in London; Galeries Lafayette, Le Bon Marché and Printemps in Paris; Rinascente in Italy; Bijenkorf in the Netherlands, KaDeWe, Karstadt, Horten, and Kaufhof in Germany; Matsuzokaya, Mitsukoshi, and Isetan in Japan; and Eaton's in Canada. Well-known American department stores include Bloomingdale's, Abraham & Straus, Filene's, Strawbridge & Clothier, John Wanamaker, Marshall Field's, Rich's, Dillard's, Bullock's, The May Company, and The Broadway.

Originally department stores dominated the center-city retail scene and later served as the anchors and magnets that made regional malls successful. However, department stores have suffered from overexpansion, leveraged buyouts, the recession, competition from the discounters, off-price merchants, and specialty stores. Several department stores have responded by spinning off specialty units of their own. Analysts have predicted gloom and doom for department stores, but consumers appreciate the variety (although it can be overwhelming) and the convenience of one-stop shopping.

Mass Merchants

The largest stores are the mass merchandise chains that sell *commodity merchandise* (standard basics) in a department store format. They have masses of stores and mass-oriented budget-to-moderate prices. Most retail consultants include Sears, Montgomery Ward, discounters such as Wal-Mart, K-Mart, and Target, off-price retailers, and factory outlets as mass merchants. The largest and most successful of these chains, such as Wal-Mart, K-Mart, and Price Club, have been dubbed "power retailers" by the fashion press. Today over 50 percent of all apparel units are puchased at mass merchants and sales per square foot are now $200 for discounters compared to $150 for department stores.[11] Due to their ability to buy in volume, mass merchants are able to buy at lower prices and pass the savings on to the consumer. Most of these stores are general merchandise retailers, but we limit our discussion to those that carry fashion apparel and accessories.

Discounters

Value has become the watchword of the nineties. Even upscale customers are going to discounters, or budget retailers. These stores have more than doubled since 1980 and are taking business away from the traditional department and specialty stores. Led by Wal-Mart, discounters have become the largest retailers in the United States.

The term *discount store* developed after World War II, when certain stores claimed they could sell merchandise for less because they had lower overhead, that is, lower operating costs: low rent, less advertising, fewer salespeople, and few customer services. Today, however, stores such as Wal-Mart are able to offer as much service as any department store.

Off-price retailers

Some retailers specialize in selling merchandise at lower prices by offering special buys, closeouts, overruns (manufacturers' overproduced merchandise), last season's goods, off-colors, and manufacturers' returns. The need for low rents excludes them from high-rent malls and prime downtown areas. Examples of off-price retailers are Ross, Loehmann's, and Marshalls.

Retail outlet stores

Since manufacturers' outlet stores (see Chapter 12) and other off-price stores have been gaining market share, many department and specialty stores are attempting to reclaim lost customers by opening their own outlet stores. They feel that it is more profitable to run their own clearance stores

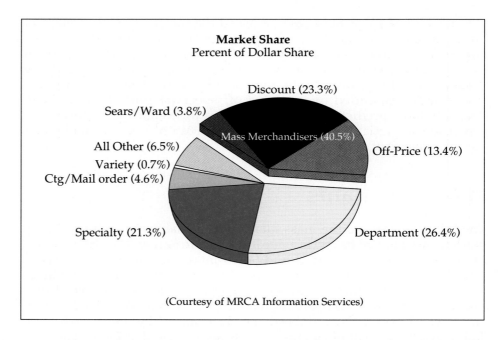

Market Share
Percent of Dollar Share

Discount (23.3%)

Sears/Ward (3.8%)

Mass Merchandisers (40.5%)

All Other (6.5%)

Off-Price (13.4%)

Variety (0.7%)

Ctg/Mail order (4.6%)

Specialty (21.3%)

Department (26.4%)

(Courtesy of MRCA Information Services)

than to sell leftover markdowns to jobbers. Macy's, Neiman-Marcus, Nordstrom, and John Wanamaker are examples of stores that have their own clearance centers.

Warehouse Clubs

Warehouse clubs, which charge a small membership fee, offer consumers deep discounts on bulk packaging. While they do offer some clothing, they are primarily general merchandise stores that offer no-frills savings in a warehouse setting. With stores ranging from 40,000 square feet to 160,000 square feet, these retailers offer approximately 3000 to 5000 deeply stocked products. Sam's Clubs, owned by Wal-Mart, is the largest with 266 stores in 42 states. Other clubs include: Price Club, Costco, K-mart's Pace, and B. J.'s Wholesale Club.

Category Killers

The term *category killers* has been used in the fashion press (although it does not usually apply to stores with fashion merchandise) to refer to stores that offer an extremely wide assortment of merchandise in a specialized field. In fashion retailing, The Gap and Kids R Us might qualify as category killers.

Destination Stores

The term *destination store* simply signifies a retail establishment that consumers seek out because they offer something special. Whether the special something is a fashion look, low prices, or service, it must be worth the trip just to shop at that store. A destination store can be a specialty store, a department store, or a mass merchant; very often it is a category killer. Current fashion destination stores include: outlet malls, Wal-Mart, and anyone's favorite store.

Promotional Stores

Any store, be it a specialty store, department store, or mass merchant, that is price directed is a promotional store. *Promotional stores* offer special buys from manufacturers and frequent sales in an effort to get customers into the store. Most consumers like to buy on sale because they feel they are getting a bargain. Examples of promotional stores are Macy's, Mervyn's, and Target.

Non-store Retailing

Non-store retailing is posing a very serious threat to the traditional retail industry and to the 19 million people it employs. Another form of *direct retailing*, it eliminates the need for middlemen and buyers.

Many people dislike the inconvenience of traffic, parking, crowds, and going from store to store in search of a coordinating wardrobe. Busy people do not have the time to go shopping in a store. They want to buy clothing and accessories quickly and efficiently.

Mail order

Mail order is answering the needs of the person who dislikes shopping. The number of consumers shopping by catalog has almost doubled in the past decade.[12] It is easy to order from catalogs at home and to have the merchandise delivered. Catalogs provide an opportunity to compare merchandise and prices while sitting at home. Apparel catalog retailers include Spiegel, Tweeds, Bachrach, Land's End, L. L. Bean, and J. Crew.

Started in 1983, J. Crew averages about 200 items per catalog. In 1992 it mailed 72 million catalogs. Arthur Cinader, chairman, explains that "we would be committed to 70 percent of a planned need for an item before the first catalog is in the mail. For example, on a $60 item, we might project $300,000 in volume, and have ordered 3500 units before there's any demand at all."[13] The firm gets a 4 to 5 percent response rate and takes in an average order of about $110, or $5 per catalog.

Roger Horchow, formerly a mail-order executive at Neiman-Marcus, started his own profitable luxury-item catalog business in 1972, only to have it be purchased by Neiman-Marcus in the late 1980s! Unfortunately, Sears closed its venerable mail-order business in 1993 because their catalog had become too broad, poorly managed and merchandised, and therefore, unable to compete successfully.

Many mail-order businesses, including Talbots and J. C. Penney, also have retail stores. Some added stores after their catalog businesses became successful; others were already established retailers. In fact, all retailers are finding that mail order brings them increased business. Bloomingdale's used to put out catalogs just to draw customers into its stores; now mail and phone orders account for a substantial part of its business. Mail order is a difficult business, however. Catalogs are expensive to prepare and distribute, postal rates have gone up, and sales tax must be figured for many states. Only the best and most focused of the mail-order retailers will survive.

Party-plan retailing

Often overlooked by the industry, at-home party retailing is a growing vehicle for direct sales of women's wear, children's wear, and jewelry.

According to Washington D. C.'s Direct Selling Association, selling products in the home has grown to a $10 billion industry. Sales consultants (90 percent of whom are women) encourage prospective customers to invite friends into their homes to look at and try on merchandise. The pricey Carlisle Collection, for example, is sold through a nationwide network of at-home consultants.

Electronic retailing

A number of on-line computer information services now offer a shop-by-computer feature. On-line services that require a personal computer (PC), a modem, and a computer-savvy consumer are few but growing. The largest providers of these services are CompuServe, Prodigy, GEnie and American Online. On the Prodigy system, buyers can purchase from more than 50,000 entries, including clothes from Spiegel or J. C. Penney. One hundred merchants that sell on CompuServe are sending color catalogs through the telephone lines.

Brownstone Studio, a mail-order firm, is experimenting with video catalogs. Some stores are also testing video order machines similar to bank automated tellers outside their stores. When the store is closed, a customer can view merchandise on a video screen outside the store, and using a charge card, order the merchandise and have it delivered.

Behind-the-scenes at Home Shopping Network.
(Courtesy of HSN)

Cable television retailing

Fashion selling on cable television is claiming national attention. Television allows consumers across the country to see fashion in their own living rooms, order it by phone, and have it delivered. As with catalog shopping, the drawback is that the clothes can not be touched or tried on. Home Shopping Network (HSN) emphasizes low prices while QVC (Quality, Value, and Convenience) uses a talk-show format featuring product information. Saks Fifth Avenue, Arnold Scassi, Diane von Furstenberg, and Liz Claiborne have enhanced QVC's image while Macy's intends to start its own television channel. In the next few years it is expected that television home shopping will be done through a remote-control device.

MULTIPLE-UNIT STORES

Retailers begin as single-unit stores and, if successful, may branch out into multiple units.

Stores that carry on the retail tradition may be classified as *single-unit* (one store only) or *multiple-unit* (more than one store) operations. These range from small, single stores with just a few employees to giant retail firms with thousands of employees. Department stores, specialty stores, and price-directed stores can be found in either classification.

The success of the multiple-unit store is based on its ability to buy in large quantities and to distribute the merchandise and operating costs throughout the organization. There are two types of multiple-unit organizations: chain stores and groups.

Chain Stores

The stores in a chain organization are basically uniform. All are merchandised from a central office, often in or near the *flagship store*, the largest and most representative store in the chain. Many regional and national specialty stores, such as The Limited, Ann Taylor, Saks Fifth Avenue, and I. Magnin, as well as mass merchants like Sears and J. C. Penney, are chain organizations. Because the term *chain* has been associated with low-priced merchandise in the past, many specialty stores simply call themselves *national specialty stores,* even though they are structured as chains. Some chains have become megaretailers. J. C. Penney has 1360 stores, and Wal-Mart and K-mart each have over 2000 stores in the United States. An international example is U. K.-based Laura Ashley, one of the first in global retailing, with 450 shops worldwide.

Although most buying is done centrally, allowances are made for regional differences. A store in Connecticut, for example, would need more heavy outerwear in winter than a store in Pasadena, California. Centrally purchased merchandise is distributed to all stores from a central or regional distribution center (see Chapter 14).

The Limited Express flagship store, on Madison Avenuen New York City, is part of a nation wide chain store organization. *(Courtesy of The Limited)*

Department Store Groups

Many multiple-unit department stores use the term *groups* to describe their organizational structure of a parent store with branches. Several examples include Bloomingdales, John Wanamaker, and the May Company. These department stores originated years ago in metropolitan areas and later opened branches in outlying suburbs to serve customers who enjoyed the convenience of shopping closer to home.

Store groups are also centrally merchandised, usually at the parent store. The parent or flagship store is usually the focal point of the group, carrying a wider assortment of merchandise than the branch stores. In some cases, however, popular branches now sell more volume than the flagship.

Franchising

Although a rarity in retailing, some manufacturers sell the rights to retail their merchandise. Manufacturers benefit from this arrangement because the product must be sold under the brand name and merchandised according to their specifications, which protect the manufacturer's image. Retailers benefit because they are guaranteed availability of stock and the right to use the brand name in advertising which is supported by the manufacturer's national or international advertising. The store has no rights to selection from the line but must carry whatever merchandise is sent to them. Examples of international franchises are Mondi and some Hermès stores.

STORES WITHIN A STORE

The intimate atmosphere of small shops can be created within a larger store.

In-store Designer Boutiques

Popular designers or brands require that a retail store furnish a permanent location and *real estate* (square footage) within the store for the sole purpose of displaying only that designer's merchandise. The designer or brand vendor prescribes how the merchandise is visually merchandised and accessorized. For this reason, it has become common for the vendors to supply the fixtures. For example, there is the Ralph Lauren Polo boutique with its mahogany shelves, fireplace, polo sticks, umbrella stands, shaving mirrors, and other "British gentleman" accessories. Many of these same designers and manufacturers also have their own freestanding shops, discussed under Specialty Stores, which compete with these in-store boutiques. Wal-Mart has even opened a "vendor store" where manufacturers are permitted to take charge of selecting and displaying their own merchandise.

This merchandising strategy does not particularly help a store to maintain a cohesive image, except as a store carrying designer names. Retail consultant Walter Levy complained, "When I worked at

The Donna Karan shop at Saks Fifth Avenue. *(Courtesy of Saks Fifth Avenue)*

Bloomingdale's many years back, we had a storewide point of view. Today, it's a series of vendor names along the wall."[14]

Leased Departments

To provide additional services for their customers, many stores lease certain departments to an outside organization better able to handle a particular specialty. Leased departments in retail stores are merchandised, owned, and operated by an outside firm rather than by the store itself. Departments such as furs, shoes, and fine jewelry are sometimes leased because special training for sales associates is necessary. Today leasing is used less frequently, however, as it is more cost effective for a store to run its own departments.

RETAIL ORGANIZATION

There are a number of career opportunities within each area of retail responsibility.

Retailing functions are generally divided into six areas of responsibility: marketing, merchandising, store operations, sales promotion, finance and control and personnel.

1. **Merchandising** has the responsibility for planning, buying, and selling merchandise (see Chapter 14).

2. **Store operations** maintains the retailer building, protects the store and merchandise, provides customer services, and coordinates the movement of goods and people within the building.

3. **Marketing** directs the focus and image of the store to provide the store's target customer with the right merchandise at the right time and to make a profit doing it. Marketing also includes the responsibility for sales promotion.

4. **Sales promotion** informs customers about goods and services through advertisements, displays, publicity, special events, and public relations (see Chapter 15).

5. **Finance and control** keeps records of money spent and received: accounts payable (goods received but not yet paid for), payroll, taxes, credit (customers' charge accounts), and inventory.

6. **Personnel** staffs the store with people who are qualified and trained to handle the work that needs to be done and assures compliance with state and federal labor laws.

The Small Store

In a small store, all the retailing functions may be carried out by the manager and a few assistants. Small stores can offer their customers highly personalized service because of the opportunity for frequent contact.

The typical small store is often managed by one person, usually the owner, who assumes many roles in the store's operation such as manager, buyer, salesperson, stockperson, and bookkeeper. In some cases, several people jointly run a store, each assuming the responsibility for the jobs he or she is most capable of doing. Problems occur, however, when the people involved have different goals or inadequate experience. Many small retailers in the United States go out of business because of inadequate financing, poor financial management, and competition.

Joseph Cicio, CEO of I. Magnin, discusses marketing with Carl Portale, publisher of Harper's Bazaar, at a reception honoring Harper's Bazaar editor Liz Tilberis. (*Courtesy of I. Magnin*)

The Large Store

In large stores, the retailing functions are divided among people who are specialists in the performance of their respective jobs. The function of the management team is to set policies and make sure that strategies are being carried out properly. Depending on the ownership, there is usually a chief executive officer (CEO) to oversee day-to-day operations. Several top-level executives are usually in charge of the various functional responsibilities such as merchandising, operations (store directors), sales promotion, personnel, and finances. The larger the store—by sales, square footage, or number of stores—the more complex the organizational structure.

MARKETING

Retail management determines its target market and tailors the store's image and merchandise focus to answer the needs of their customers.

Retail management must decide which potential customers it wants to reach and how it wants to reach them. Whether setting up a new store or reevaluating an existing one, management must study the type of people who live in the community, their life styles, and ultimately their shopping wants and needs. It must then determine how to fill these wants and needs.

Retail Target Customers

As in design and manufacturing, the first step in reaching potential customers is to define exactly who they are. No retail store can be all things to all people; it must select one or a few groups of people to serve. *Target market* is the term for the group of consumers the store wishes to attract. Massimo Ferragamo explained, "Today, you can't take the customer for granted and just hope they come floating into the store. We want to know who they are, their likes and their dislikes."[15] A group of target customers are within a general age range, have similar life styles, and therefore have similar needs and tastes. A department store may try to appeal to several categories of target customers. Separate areas or floors are created for each category; for example, designer, contemporary, or junior.

Store location

The location of a store is very important in relation to its potential customers. *Demographic research* is done to determine what location would best attract target customers. Computerized site-selection programs provide data on area population forecasts, descriptions of households by income, median age of market-area residents, and information about the competition. However, since time constraints have caused people to shop less, retail space is expected to decline rather than grow during the next decade.

Renovations

Rather than building new stores or expanding, many retailers are using capital investments to renovate, strengthening existing units to compete successfully. At the National Retail Federation convention, Michael Gould, Chairman and CEO of Bloomingdale's, and Sally Frame Kasaks, Chairman and CEO of Ann Taylor, were among panelists who concluded that renovation is a critical way to reposition a store, generate new traffic, and broaden the customer base without the expense of building a new store.[16]

Fashion Leadership

In conjunction with the target customer, the store must determine what role it wishes to play in fashion leadership. Fashion leadership can be separated into three loosely defined categories:

1. **Fashion-forward stores,** such as Bergdorf Goodman, Saks Fifth Avenue, I. Magnin, or Neiman Marcus, seek leadership by carrying fashionable merchandise. There are fewer of these stores because the percentage of customers who can afford or have an interest in fashion newness is relatively small.

2. **Main-stream retailers,** whose fashion image falls in the moderate fashion and price category, identify with consumers who accept popular fashion. Fashion adaptations are available to them at moderate and upper-moderate prices. Most of the department stores, such as Macy's or Dillard's, fall into this category.

3. **Mass merchants** such as Wal-Mart and Sears restrict their fashion offerings to proven styles because they appeal to fashion followers and to people who simply cannot or will not spend more money on their clothing. Mass merchants feel that, for their customer, fashion is not the issue. Bob Connolly, at Wal-Mart, commented that, "Opulent runway merchandise was never intended for the majority of people... ."[17]

The Retailing Image

It is important for a retailer to clearly define its *image*, the personality or character that it presents to the public. This image or uniqueness reflects its degree of fashion leadership and its market niche and therefore appeals to its target customers. Ralph Lauren says, "Retailers have to have a point of view. It's the most important thing for ... retailers ... to have an identity, a sense of who they are... ."[18]

In keeping with that image, retailers strive for a store atmosphere that is both a complementary background for merchandise and an inviting environment for the customer. The store's merchandising, interior decoration, promotion, and customer services develop, maintain, and reflect that image and try to generate excitement about the merchandise in order to create a desire to buy.

SUMMARY

This chapter examined the current retailing situation: the closure and restructuring of stores caused by debt and fierce competition. Retailers are trying to cope by meeting consumer needs and offering value at reasonable prices. Many retailers feel that increased market share can only come from global expansion. There is a renewed interest in inner-city retailing and malls are diversifying to stay in business.

There are many types of retail operations including specialty stores, department stores, mass merchants, mail order, and electronic retailers. There are both single-unit and multiple-unit stores, some growing into large chains. Store functions include marketing, merchandising, store operations, sales promotion, finance, and personnel. In small stores all of these jobs are handled by only a few people; in large stores, each function is directed by a different executive.

In order to maintain a competitive position, a retailer must be focused on the needs of its target customer. To appeal to that customer, the store must have a unique yet appropriate image carried out in its merchandise, promotion, and services.

CHAPTER REVIEW

Terms and Concepts

Briefly identify and discuss the following terms and concepts:

1. Over-stored
2. Leveraged buyouts
3. Fifth Avenue
4. Focused merchandising
5. Specialty stores
6. Deep-niche retailing
7. Vertical retailers

8. Off-price
9. Leased departments
10. Franchising
11. Electronic retailing
12. Marketing
13. Vendors
14. Target customers

Questions for Review

1. What is the purpose of retailing?
2. How does globalization affect retailing today?
3. How does a department store differ from a specialty store?
4. Explain the differences in organization between a small store and a large store.

5. Why has off-price, discount, and warehouse retailing increased over the last few years?
6. Why is a store's image important?

Projects for Additional Learning

1. Find out whether one of the department or specialty stores in your community is owned by a retail corporation. Find out what other stores are part of the same corporation (see Appendix Two).
2. Research the history and growth of a large store in your area. Contact the store's public relations department for information and check your college or local library for further information.

3. Watch a fashion presentation on the QVC cable network. Write an analysis of the program including the effectiveness of the moderator, the presenter, and the merchandise. Was the show successful in terms of sales? What do you think of the future of television shopping?

NOTES

[1] "Modes of Retailing," Merrill Lynch Seminar, March 1992.

[2] Interview, April 29, 1992.

[3] Bob Connolly, Wal-Mart, letter, May 27, 1993.

[4] Howard Rudnitsky, "Battle of the Malls," *Forbes*, March 30, 1992, p. 46.

[5] Interview, April 29, 1992.

[6] As quoted by Barbara Solomon, "Malls Mix it Up," *Women's Wear Daily*, March 17, 1993, p. 7.

[7] John Mannix, U. S. Mall Manager for Sarakreek, owner of South Hills, as quoted in "Battle of the Malls."

[8] Interview, March, 1993

[9] "Upscale Boutiques," *Women's Wear Daily*, March 4, 1993, p. 8.

[10] As quoted by Phyllis Berman, "Closer to the Consumer," *Forbes*, January 20, 1992, p. 57.

[11] MRCA Information Services as quoted by Ira Schneiderman, "Mass Market Report," *Women's Wear Daily*, February 24, 1993, p. 22.

[12] Gretchen Morgenson, "The Fall of the Mall," *Forbes*, May 24, 1993, p. 108.

[13] As quoted in a letter from Gerry Hammarth, J. Crew, Inc., May 6, 1993.

[14] Walter K. Levy, Retail Consultant, speaker, The Fashion Group Annual Retail Forum, January 19, 1989.

[15] As quoted by Chuck Struensee, "Ferragamo Today," *Women's Wear Daily*, February 11, 1991, p. 7.

[16] As quoted by Dianne Pogoda, "Spruce Up to Stir Sales," *Women's Wear Daily*, February 1, 1993, p. 8.

[17] Connolly, May 27, 1993.

[18] As quoted in "Lauren at 25," *Women's Wear Daily*, January 15, 1992, p. 7.

Retail selling floor of Gieves & Hawkes, Savile Row, London. *(Courtesy of Gieves & Hawkes)*

RETAIL FASHION MERCHANDISING

CAREER FOCUS

Fashion merchandising is directed by store management, merchandise managers, and sometimes the fashion director. However, the actual buying and sell-through of merchandise is the responsibility of buyers and their assistants. Buyers are also involved in sales promotion as discussed in Chapter 15.

CHAPTER OBJECTIVES

After reading this chapter you should have attained competence in the following areas:
1. Awareness of the buying-selling aspects of merchandising.
2. Understanding of the importance of planning.
3. Ability to explain exclusivity versus national brands.
4. Understanding of buying procedures at the market.
5. Comprehension of all aspects of inventory control.
6. Ability to explain the services offered by buying offices and the difference between a buying office and a store buyer.
7. Understanding of the importance of customer service and sales associates.
8. Understanding of the role computers play in retailing.

Merchandise is the term used to signify articles for sale; it derives from the word *merchant*, the actual seller or retailer. *Fashion merchandising* includes all the planning and activities necessary to supply the fashion wants and needs of retail customers. In the past, fashion merchandising was usually associated only with women's apparel and accessories. Today, however, stores use aggressive merchandising techniques for men's and children's clothing as well. Fashion influence has also spread to other areas of retailing, from home furnishings to cookware.

This chapter will cover the planning and carrying out of buying and selling, including the responsibilities of the buyer. We will follow the flow of merchandise from arrival in the store to purchase by the consumer.

MERCHANDISING ORGANIZATION

Every area of merchandising responsibility needs planning and organization to make it function properly and to ensure successful buying and selling.

Usually merchandising responsibilities are divided between two chains of command. The *buying line* of management has responsibility for merchandise content and assortment; the *store line* is responsible for operations. The goal of both is to sell merchandise.

The Responsibilities of Operations Management

The main responsibility of operations management is to maintain the retail building, protect the store and the merchandise, provide customer services, and coordinate the movement of goods and people within the building. Above all, they must work together with merchandisers to produce positive sales results. Everything and everyone involved in the operation of the store must be organized to achieve this goal. The director of stores, an executive vice-president or general manager, has the responsibility for the store line. The director of stores supervises individual store managers in multiple-unit organizations and they, in turn, delegate responsibilities to group sales managers, floor or area managers, who in turn supervise department managers. Department managers and their assistants run the department and supervise sales associates.

The Responsibilities of Merchandisers

The buying line must do all the planning and other activities necessary to bring the right merchandise into the store at the right time to satisfy the store's customers. Responsibility for sales differs among specialty, chain, and department stores.

Divisions

Merchandising responsibility is segmented into *divisions* such as women's better sportswear and men's furnishings. Several divisions are grouped together under the direction of a general merchandise manager (GMM) who reports directly to the chief executive officer. A GMM for women's ready-to-wear, for instance, might direct several divisional merchandise managers who are in charge of missy dresses, missy sportswear, junior dresses, junior sportswear, and so on. Otherwise, merchandise divisions may simply be grouped together by floor; all the merchandise on one floor of the store may be under one divisional merchandise manager.

Departments

Each division is composed of *departments*. For example, accessories is subdivided into handbags, hosiery, hats, jewelry, and so on. Each department is further segmented into *classifications*. Missy dresses might be divided into better, moderate, and budget price groupings. A buyer is responsible for the success, measured in profits, of one classification or department.

Fashion Merchandising Direction

Fashion direction is established to maintain cohesive fashion merchandising in line with a distinctive store image. In single-unit stores, the owner usually acts as fashion director and buyer. In large chains, management may give direction, as Gene Pressman does for Barney's, or management may employ a fashion director, such as Kal Ruttenstein at Bloomingdale's, who is involved in marketing, merchandising, and promotion. The fashion director is the bridge between corporate marketing policy and actual merchandise-buying decisions, and major decisions may be made as a team. He or she works with merchandise managers, buyers, and promotion executives to suggest what merchandise to choose and how to present it.

Along with management and the designer collection buyers, the fashion director attends European and American collection openings to study fashion trends, relates those trends to the store's image, and passes this information on to other buyers as a guide to merchandise planning. The fashion director may also work with buyers to select appropriate merchandise, to develop the store's private label, and to coordinate their buys with merchandise in other departments. A fashion director also prepares seasonal fashion presentations for sales associates so that they can understand the new fashion concepts and the store's merchandising approach and better help their customers.

Buyers and management determine a *fashion merchandising policy*, a long-range standard for fashion buying and selling, and related activities such as promotion.

BUYING PREPARATIONS

Careful planning is done to help merchandisers do efficient and successful buying.

The Merchandise Plan

The *buyer* makes up the merchandise plan within the framework of the policy, goals, and fashion direction set by management. The *merchandise plan*, expressed in dollars rather than units, is a financial plan allocating specific amounts of money to each department or division for the purchase of an appropriate assortment of fashion merchandise that will meet consumer demand. Merchandise plans are determined four to six months before the selling season and cover a six-month period: the spring season (February through July) or the fall season (August through January).

The buyer uses actual sales figures from the corresponding season of the previous year as a basis for making up the new plan. The buyer develops the plan on computer spread sheets which show what needs to be purchased and sold per month to reach sales and profit goals. As the planning of merchandise has become increasingly complex, the computer has helped to organize facts and figures that would otherwise be overwhelming. The plan includes anticipated sales, stock, markdowns, and purchase plans for the coming six-month season. As the only source of income for most retail stores is the sale of merchandise, the projected sales must cover all expenditures such as the cost of wholesale merchandise and operating expenses as well as *gross margin* or profit.

Planning sales goals

To make a realistic estimate of prospective sales, a buyer must consider variations in consumer demand, shifts in population, local retail competition, physical expansion or alterations needed in the store, planned promotional efforts, market and trend analyses, seasonal consumer demand, and economic conditions. For instance, buyers tend to buy short when anticipating a recession. In a good year, they buy more in hopes of an increase in business.

Planning stock

The next step in planning is to determine the amount of stock, in terms of dollar investment, necessary to meet consumer demand and thereby support planned sales. The same influences that affect sales also affect the planning of stock. Stock must be brought to a peak just before the expected time of peak selling.

A high and quick stock-turn rate is the desired result of stock planning. *Stock turn* is the number of times that inventory (merchandise stock) has been sold and replaced during a given period. The quicker stock is turned, the more income and profits the store makes.

Planning purchases

The buyer determines the amount of merchandise, in terms of dollars, that may be purchased during a given period. Considering stock on hand at the beginning of any given month, the buyer has to calculate the amount of purchases that can be made if stock and sales are to be kept in balance. The difference between actual stock and planned stock equals *open-to-buy*, the value of planned purchases.

The Buying Plan

By stating sales objectives and expenses, the merchandising plan provides the foundation for assortment planning. A *merchandise assortment* is a collection of various styles, quantities, and prices of related merchandise, usually grouped under one classification within a department. The buyer plans to buy a balanced assortment of merchandise to meet consumer demand and appeal to a particular group of target customers.

Assortment planning is expressed in the form of a *buying plan,* a description of the types, quantities, prices, and sizes of merchandise that a buyer expects to purchase within a specific period of time. The totals state exactly how much may be spent on merchandise in each category in line with sales goals and the dollar merchandise plan.

The more detailed the buying plan, the less confusing buying decisions will be, allowing the buyer to concentrate on the fashion aspects of the merchandise during the seasonal market. The plan must be flexible enough, however, to allow for revision if conditions change.

BUYING

The buyer purchases merchandise in accordance with the merchandising plan and sales and profit goals.

The Buyer's Role

The buyer tries to select merchandise from vendors that the store's customers want or need: the right styles, size range, color assortment, and fabrics, all at

Buyer Julia Ellard, social occasion dresses buyer for I. Magnin, and her assistant Kim Baur at work in their office. *(Courtesy of I. Magnin, photographed by the author)*

acceptable prices. The buying process is part analytical and part creative. The mechanics involve knowledge of sales histories and development of a merchandising plan. The creative side is the ability to understand the customer, spot trends, and use intuition to choose merchandise with terrific sell-through. Some retailers, such as Macy's, are experimenting with separating these functions into a *buyer-planner system*. Under this system, buyers focus on shopping the market and merchandise selection, while planners shape the size of the buy and how it's to be distributed.

Individual preferences must be forgotten in favor of the buyer's knowledge of customer preferences. Buyers need to keep in touch with their customers' life styles in order to buy merchandise to fit their needs. One San Francisco-based buyer of long formals and evening dresses spent a late spring evening sitting in the lobby of a downtown hotel to see what young people's preferences were in prom dresses.

A buyer's knowledge of merchandise stems from both education and experience. The ability to evaluate merchandise and judge whether it is suitable for a customer develops over years of examining all types of merchandise for quality, styling, and price. Leslie Wexner, founder of The Limited, said that his "experience of being in stores, watching customers' reactions to merchandise and its presentation, and overhearing their comments," helped him develop a retail sense.[1] Market and trend research becomes second nature to the buyer. In both planning and buying, the following factors are considered:

- ◆ The store's fashion image and merchandising policies
- ◆ Market and fashion trends
- ◆ The effect of economic conditions on demand for certain types and prices of merchandise
- ◆ Basic stock—merchandise that is in consistent demand throughout the year or the same season each year
- ◆ The competitors' merchandise offerings
- ◆ The individual department's ability to house and display the merchandise effectively.
- ◆ The type of promotional activities that are needed to support the merchandise

Unfortunately, many buyers have become isolated from their customers with offices removed from the selling floor. In a large chain a buyer cannot visit every store. Their increased responsibilities to the bottom line leave many buyers no choice but to be just administrators. As retailers explain in industry jargon, they have had to become "number driven" instead of "merchandise driven." Several stores such as Nordstrom and Barney's, however, insist that their buyers spend time on the selling floor.

Corporate Buying

In an effort to be efficient, buying between large retailers and large manufacturers is often done management to management. This is often done to save the money that would be given to sales commissions. Wal-Mart, for example, tries to do all its buying in this manner. In this case, the buyer is part of a buying team.

The Buying-Selling Cycle

A buyer's responsibilities include both the buying and selling aspects of retailing. Therefore, the job involves a complete cycle: planning what to buy; searching the markets and selecting the right merchandise; working with advertising, display, and special events to promote merchandise and motivate consumers to buy; organizing and arranging the merchandise in the departments; and training personnel in sales—only to begin the cycle again by evaluating those sales in order to plan for the next season. Thus, the buyer works in two time zones, anticipating future needs and evaluating current sales. In fact, because consumer tastes change constantly, the buyer's job of identifying and interpreting consumer demand is a continuing process.

   The buying and selling cycle is related to the fashion cycle of consumer acceptance. Ideally, the buyer would like to buy a broad but shallow assortment of merchandise at the beginning of a season to test consumer reaction and then, as certain styles emerge as best-sellers, to increase stock in depth. However, most popular lines are sold up quickly and do not accept reorders. The other preference is to buy very close to the selling season (short-cycle buying) to be better able to judge market conditions and trends and to avoid markdowns. As the peak selling period passes, leftover merchandise is marked down for clearance and buying begins again for a new season. The buying-selling cycle is constantly overlapping: new goods come into the store while other goods reach their peak in sales or decline in sales.

Shopping the Market

After the buying plan has been established, fashion buyers shop the market to view the merchandise available for the coming season. At this point the buyer meets the manufacturer's representative. For the retailer, the manufacturer is the *supplier, vendor,* or *resource* of fashion goods.

   Buying trips are generally timed to cover markets that are important for a particular category of merchandise. Buyers visit different market centers for different needs. Many people imagine a buyer's job to be a glamorous one involving many trips to Europe. However, only designer-department buyers, merchandise managers, and fashion directors of large stores attend the European collections. Even then, with packed appointment schedules leaving little time to eat or relax, business travel does not remain glamorous for long. Many French and Italian designers now have New York showrooms so that buyers do not necessarily have to go to Europe.

   Designer and contemporary buyers go to New York to see the American collection openings. Contemporary and junior sportswear and dress

Buyers from all over the world at the Salon du Prêt-à-Porter Féminin in Paris.
(Courtesy of Profem)

buyers also check out the California markets. To find moderate and budget-priced apparel, buyers may go to manufacturers' showrooms at regional markets. Buyers from Dillard's, for example, are buying half of their merchandise at the Dallas markets. Merchandise may also be bought from manufacturers' representatives who call on stores.

Line buying versus trend buying

The buyer shops for new fashion from both key resources and new ones. *Key resources* are vendors who have maintained a reputation for dependability and whose merchandise sells through because of appropriate styling, quality, and price, such as Liz Claiborne. The buyer regularly buys a good portion of these manufacturers' lines, a practice called *line buying.* However, the buyer is also always on the lookout for new resources and new talent. Finding an exciting, unique resource can mean an important merchandising statement for a fashion store. However, because buyers are spending more and more time administering and have less time for actual buying, they are often unable to look for new resources. New designers and manufacturers must strive to get the attention of buyers. Buying merchandise for its innovative styling is referred to as *trend buying.* Generally, a combination of line buying and trend buying is implemented.

Shopping procedures

Most buyers visit several showrooms of manufacturers who produce the specific types and price lines of merchandise they need. As the entire line is shown to them by management or a sales representative, they take the numbers of the styles they like, along with careful descriptions and notes on available size and color ranges, fabrication, and wholesale price. In the case of a major store making a large purchase, buying is likely to be done by a group of executives including the buyer. The designer and/or merchandiser and sales manager present the line, making recommendations. British designer Jasper Conran tells of one buyer who asks him, "'Jasper, what do *you* think is best?' She makes you edit your own collection. And since you always do know what's best, she gets the core."[2]

Buyers must evaluate whether styling and quality compare favorably with other merchandise they have seen in the same price line. They must also consider the possible *sales potential* of each style for their store. Sales representatives try to be helpful, pointing out successes in the line based on initial orders. At the end of the market trip, a buyer compares his or her notes on merchandise seen, eliminates the less desirable styles and any duplications, and decides which styles to purchase.

The buyer as editor

Although retailers are not usually fashion creators, they can influence consumer buying to some degree by their preselection of merchandise, which narrows the choice for the ultimate consumer. By determining what parts of a collection are bought, and in what quantity, the buyer affects not only the fortunes of a designer's company but also the public perception of the designer's entire line. In this sense the buyer is a fashion editor, which is one of the main reasons many designers are opening their own stores.

Purchase Orders

Placing an order for merchandise is considered a contract between the store and the vendor. Therefore, writing an order commits the store to taking the merchandise, if it meets quality expectations and delivery requirements. Generally, the buyer writes up orders on the store's purchase order forms and has them countersigned by the merchandise manager. Purchase orders specify the date of the order, the name and address of the resource, the terms of sale, shipping instructions, the store's shipping address, the name of the department, the quantity ordered, descriptions and prices of styles ordered, and obligations between buyer and seller. Purchase orders are most efficiently done by computer linked with the vendor, which can instantly supply information on what goods are available or are in work and what shipping dates are expected.

Deliveries are timed so that sufficient quantities of merchandise are in the store to meet various peaks in the consumer demand cycle. The vendor is committed to meeting those delivery dates or the order may be cancelled.

Terms of sale

As indicated on the purchase order, *terms of sale* specify how soon after shipment the invoice must be paid. Manufacturers offer cash discounts to retailers if payment is made within this specified period. Discounts are given to encourage prompt payment but the retailers have come to expect it even when late. The cash discount in the women's apparel industry is 8 percent if the invoice is paid within 10 days of the end of the billing month.

Computer Use in Buying

In an effort to cut down reaction time in response to consumer demand, retailers are using a strategy called *Quick Response,* which utilizes integrated computer systems that link them to all of their stores and to certain vendors. As discussed in Chapter 12 from the vendor's point of view, this strategy attempts to speed ordering and distribution and reduce inventories via electronic data interchange (EDI) between textile and apparel producers and retailers. The goal is to move the merchandise more rapidly down the supply pipeline and feed more accurate information on consumer demand back up the pipeline.

In order to utilize EDI all goods at each level of the industry are given UPC bar codes, which identify style, color, size, price, fabrication, and vendor. Linkage systems and standardized codes have been established between vendors and retailers. The retailer's computer information systems are linked with those of the vendor by EDI in order to facilitate quick ordering and, if possible, reordering. To implement Quick Response to the fullest extent, retailers must be willing to adopt a continuous open-to-buy position, to let suppliers replenish without any retail management approval or review.

Quick Response requires a large capital investment to implement UPC and EDI technology. However, retailers feel the cost is justified and gives them a competitive edge. Early proponents of QR included Dillard's, J. C. Penney, Strawbridge & Clothier, and Wal-Mart. Federated has implemented a sophisticated new computer system with a program called FASST (Federated Accelerated Sales & Stock Turn) whereby suppliers keep each store stocked at all times.

Buying Offices

To facilitate buying, many stores are affiliated with resident buying offices (located in international and domestic market centers) or have their own corporate buying offices. Buying offices are not a substitute for buyers, as the buyers still do the actual buying, but instead act as market representatives for them. Because they are a daily presence in the market, they can do the market research, saving retailers time, money, and effort.

Buying offices, too, have been faced with mergers and acquisitions. Lifelong competitors have found themselves facing one another across boardroom tables. The two primary types of buying offices are independent and store owned.

Independent resident buying offices

Independently owned and operated, these buying offices charge fees to noncompeting stores for market services. The current largest is The Doneger Group which represents over 600 stores.

Store-owned resident buying offices

There are basically two types of store-owned buying offices: associated and corporate.

An associated buying office is jointly owned and operated by a group of stores. Member stores usually have similar sales volume, store policies,

An illustration from the Frederick Atkins *Trend Direction* brochure.
(Courtesy of Frederick Atkins, Inc.)

and target customers but are in noncompeting locations. Operating expenses are allocated to each member store on the basis of the store's sales volume and the amount of services rendered. These offices may also charge a fee for limited services to nonmember stores. Associated Merchandising Corporation (AMC) and Frederick Atkins are well-known examples.

A corporate buying office is owned and operated by the parent organization of a group or chain of stores. At Federated, 70 percent of the buying for member stores is done centrally by the parent company while the other 30 percent is done by individual member stores.[3]

International buying offices

Many large retail stores have their own buying offices abroad or use foreign commissionaires. *Commissionaires* are agents representing stores in foreign market centers. A commissionaire is the foreign equivalent of an American buying office. These offices are equipped to handle import/export transactions in the language of the country, check quality control, figure currency-exchange rates, provide a consolidated center for shipping, and wade through customs red tape. International buying offices, staffed by company personnel in foreign market centers, aid visiting buyers or do the actual buying themselves. Macy's, for example, has their own corporate buying offices all over the world.

Buying-office services

The buying office is organized along the same lines as a retail store. There are merchandise managers who supervise groups of market representatives. Market representatives, like buyers, are specialists who cover a narrow segment of the market. However, unlike buyers, they do not make final decisions as to purchases and do not place orders unless specifically asked by the buyer.

For member stores, market representatives see the lines of new as well as established resources, saving the store buyer trips to the market; send out bulletins reporting on new fashion directions, best-sellers, trends, and special price offerings; and check into general conditions of supply and demand. In addition, buying offices can offer advice on promotions and operations based on the experience of other member stores. Buying offices also organize group purchases for small stores, so that the total order is large enough to meet the minimum-order requirements of large or important manufacturers. They may also develop private-label merchandise for their members.

Private Label

Stores across the country try to provide their customers with a wide selection of popular national brands and designer labels. *National brands* (manufacturers' brands available nationwide) help the customer identify with a consistent standard of styling, philosophy, and quality from season to season. However, as stores stock more and more national brands, exclusivity tends to decrease and stores all look the same.

Exclusivity has long been an important aspect of a fashion store's uniqueness. Buyers seek distinctive fashion looks from out-of-the-way sources in foreign markets or from little-known, aspiring young designers, to give their customers something that no other store has. Barney's boasts that almost 70 percent of its inventory is exclusive, much of it created by Barney's under its own label.

Called *private label*, this type of merchandise carries the store's label such as "Saks Fifth Avenue Real Clothes" or a fictitious name such as Macy's "Morgan Taylor," or "Charter Club." Fictitious names give the impression that the merchandise is a national brand. Retailers have merchandise made directly for them to their specifications by guaranteeing a quantity order. They may design the merchandise themselves, have it copied from a "hot" item, have it designed by a design service such as Peclers, or buy from a manufacturer that specializes in private label such as Spitalnick.

Private-label merchandise offers not only exclusivity but also the opportunity to reduce margins, savings which the retailer can pass on to the consumer. Major retailers today have hundreds of lines of private-label clothing, which account for approximately 20 percent of all men's and women's apparel sold. Macy's alone sells an estimated 50 house brands.

Pricing Retail Merchandise

Retail selling prices are based on predetermined store pricing policies and on wholesale costs.

Markup

Markup is the difference between the wholesale cost and the retail price of the merchandise. It can be figured as a percentage of retail value or calculated on the basis of cost. Most retailers calculate markup as a percentage of retail price. This method is used because expenses and profits are commonly expressed as a *percentage of net sales*, which are based on retail prices. The markup must cover operating expenses and a profit. Store expenses include salaries, sales promotion, and overhead including rent, utilities, and store maintenance. All these expenses are averaged to determine a percentage markup per item.

Retailers are trying to develop a regular-price mentality, aiming at consistent lower prices instead of having to markdown merchandise so

TABLE 14-1
Retail Pricing

Pricing of a Typical Moderately Priced Dress
(see chapters 9 and 10 for wholesale costs)

Retailer's Costs

Wholesale cost		
($61.52 less 8 percent discount for prompt payment)	$56.60	
Allowance for markdowns		
(averaged over all dresses in stock)	5.00	
Allowances for shortages and pilferage	3.00	
Salaries and benefits (averaged per garment)		
Sales staff	7.00	
Merchandising and buying staff (including expenses)	8.00	
Clerical and stock room staff		
(receiving, marking, deliveries, and other expenses)	5.00	
Advertising, display, and sales-promotion staff	8.00	
Administrative staff		
(executives, credit and accounting officers,		
including expenses)	11.00	
Employee fringe benefits	2.00	
Overhead		
(rent, insurance, utilities, cleaning, and security)	10.00	
TOTAL	$115.60	
Profits before taxes	8.40	
SELLING PRICE	**$124.00**	Retail price
	- 56.60	Cost
54% Markup	$67.40	Markup

*These are approximate figures; percentages vary depending on the kind of store. Markup percentages vary according to volume and store.

often. Mass merchants are managing this type of price policy by tightening overhead, creating lower margins with private-label merchandise, and selling merchandise in enough volume to justify their reduced margins.

Price points

Several price points are offered in each department or merchandise category. The term *price range* refers to the span between the lowest and highest price point. Merchandise of comparable quality usually falls into the same price range. Within a price range there must be enough difference between price points that variations in quality at each level are obvious to customers.

RECEIVING

Merchandise purchased by buyers is received into the store or distribution center, entered into stock, and put on the selling floor.

When buyers place orders at the market or with sales representatives, they indicate delivery dates to ensure that the merchandise is in the store at the right time to meet consumer demand. Delivery dates are usually staggered so that new merchandise is constantly on the selling floor. Computers can help buyers decide on the proper distribution of that merchandise to multiple-unit stores.

Merchandise is accepted into a central receiving location, where it is counted and checked for quality. Merchandise is ticketed according to information stated on the buyer's purchase order. The ticket includes bar code numbers for vendor, season, classification, department, and of course retail selling price. When this information is recorded in a computer, the data can be used for automatic ticket printing, record keeping and accounts payable. It speeds distribution when manufacturers *pre-ticket* merchandise using electronic data interchange.

Differences between actual shipment and the purchase order must be reconciled and a decision made whether to accept or reject partially filled orders, substitutions, or late shipments. If the order is unacceptable to the buyer for any of these reasons, it may be returned to the manufacturer or an unfilled order may be cancelled. Retailers use *chargebacks* to withhold payments to vendors to cover manufacturers' mistakes such as late shipping, missing paperwork, or incorrect assortments.

From central receiving, merchandise is sent to an individual store and on to a department. The department manager also checks and verifies both the count and the information on the price tag. If approved, the merchandise moves into a stock room or onto the selling floor.

Central receiving at Marks & Spencer in England.
(Courtesy of Marks & Spencer)

RECORD KEEPING

Retailers need to keep control of inventory, what merchandise is in stock and what has been sold so they are able to evaluate what to buy in the future.

Unit Control

The basic tool that helps retailers of all sizes maintain stock is a unit-control system. *Unit control* is a system for recording the number of units of merchandise bought, sold, in stock, or on order. Records are kept of additions to or subtractions from stock, from the time an order is placed with a manufacturer until that merchandise is sold. Information for unit control comes from purchase orders, sales records, and merchandise transfers (movement of merchandise from one store in a chain or group to another). Most stores now have electronic data-processing systems for unit control that collect and process merchandise information automatically.

The major advantage of unit-control systems is that the records enable a buyer to keep track of consumer demand. If certain styles are selling quickly, the buyer may reorder or make additional purchases; if styles are selling too slowly, the buyer may make markdown plans. Unit-control systems also give a realistic growth picture in inflationary periods, when dollar-volume increases may be misleading . Sally Frame Kasaks, president and CEO of Ann Taylor, feels that "it is more realistic to make decisions about merchandise needed per store, per week based on units rather than dollars."[4]

The major drawback of unit-control systems is that they cannot determine what styles, colors, or sizes customers wanted but could *not* find at the store. Jim Nordstrom complains that computer technology, "never tells the buyer what the customer that walked out empty-handed wanted. It only tells them what they've got that's selling best."[5] So far, this information can be learned only informally, through conversations with customers. Retailers hope to gain this information in the future via consumer surveys fed into EDI.

Inventory Control

To keep tabs on the dollar value of merchandise on hand, stores record all transactions concerning sales, purchases, markdowns, transfers, and returns. As competition continues to increase, the need for accuracy and current factual information compels more and more businesses to convert their record keeping to electronic data processing.

The accounting method used for this purpose is the *inventory control system.* Most stores use the *retail method* of inventory control, which involves figuring inventory at retail prices rather than on the basis of the wholesale cost.

Book inventory

Book inventory is an accounting of records of transactions obtained from computer data. The computer generates reports on receipts, sales, additions to and subtractions from stock, stock transfers from one store to another, and even returns by customers. When an item is sold, the point-of-sale terminal automatically feeds this information into sales records and deletes it in inventory records.

Physical inventory

Taken annually to confirm the book records and to comply with accounting regulations, a *physical inventory* is the actual item-by-item count of all merchandise on hand. In case of discrepancies between the book and the physical inventories, the physical count prevails and the inventory records must be adjusted accordingly.

Stock shortages and overages

Discrepancies between book and physical inventory control are described as *stock shortages* or *overages.* Stock shortages mean a lower physical inventory than book inventory and are due to theft, damage, or clerical error. Shortages are common, and allowances must be made in initial pricing to cover these losses. Overages indicate a higher physical inventory than book inventory, are due to clerical error, and are quite rare.

CUSTOMER SERVICE

To many consumers, and in many stores, service has become more important than fashion.

American retailers were pioneers in offering customer service. Years ago, Montgomery Ward and John Wanamaker instituted guaranteed-refund policies. Today, Marianne Millies-Lacroix at the Galeries Lafayette in Paris, feels that, "stores need more service because fashion is not as strong as it used to be."[6] The success of Nordstrom and Wal-Mart, where customer service is a priority, has paved the way for a renewed interest in improved service in every store. The service in these stores has become the yardstick by which other stores now measure their own level of customer service.

Retailers are attempting to become "customer driven" (to anticipate and focus on the needs of their customers), in fact to exceed customers' expectations. Retailers are trying to be more helpful and friendly so that customers will enjoy shopping. Retailers now offer an increasing number of services, such as:

- ◆ Greeters who welcome customers
- ◆ Special programs for regular customers
- ◆ Interactive videos that supply information on product location and special prices
- ◆ Alterations
- ◆ Free use of fax machines and telephones for customers waiting to be fitted
- ◆ Free personal shopping services
- ◆ Newsletters or "look" books
- ◆ No-question return privileges
- ◆ National credit card acceptance
- ◆ 800 telephone numbers
- ◆ Extended store hours
- ◆ Gift wrapping

◆ Convenient restrooms, comfortable places to sit, and small snack kiosks

◆ Free local delivery or free parking.

In addition to all of the services listed previously, Michael Gould, CEO of Bloomingdale's, points out, "You can talk about customer service for five hours, but the first tenet of it is to have the merchandise in stock!" [8]

RETAIL SALES

Selling techniques have become very important in trying to establish repeat business and customer loyalty.

Sales Training

Fashion selling requires special training that will give sales personnel merchandise information, confidence, and motivation. In small stores, training is informal, usually based on getting experience on the selling floor. In larger stores and chains, training is likely to be more structured. Chanel, for example, includes intense training aimed at teaching the selling staff how to establish ongoing relationships with customers to build loyalty and repeat business.

Sales meetings are held by the buyer or department manager on a daily, weekly, or monthly basis. The new merchandise is presented, sometimes on models, to show sales associates how garments should be worn and accessorized. The fashion director or buyers explain the selling features of merchandise so that the selling staff in turn can point out quality, fashion, and performance features to their customers. Buyers try to infuse enthusiasm into the sales force, hoping that it will be transmitted to the customer.

Vendors are getting much more involved in retail sales. Some manufacturers are supplying *merchandise representatives* to help train salespeople and help customers. For example Anne Klein II's representatives conduct seminars on merchandising, presentation, and selling. Vendors often supply other aids to selling such as videotapes, slide shows, or brochures.

The Department Sales Manager

In multiple-unit stores with one buyer in a central location, day-to-day operations, including responsibility for sales and the maintenance of visual merchandising, is generally left to a department manager. Department managers may request more stock of a hot item and give the buyer a projected estimate of quantities they expect to sell within a certain period. In a small store, the buyer may also function as the department manager.

Selling

Sales, after all, are the key to success in retailing. The retailer's goal is to exceed its own merchandise plan sales goals. Of course, much depends on the buyer's selection of fashion assortments and whether the merchandise is in stock to meet consumer demand.

Success in fashion selling also depends on the ability of sales personnel to convey fashion ideas. Sales associates make the important customer contacts for the store. *Personal selling* is the method involving the most customer contact. In this method, sales associates actively help customers, over the counter or on the floor, to choose merchandise suited to their tastes and needs.

Many stores such as Nordstrom and Georgiou have adopted some of the salon selling techniques formerly reserved for high-priced merchandise. Sales associates are encouraged to greet customers at the door, treat customers as guests, tell them about the store's philosophy or merchandising concepts, show them around, and ask questions about their life styles, needs, and preferences. In order to build a multiple sale, associates are ready with alternative selections, wardrobe extenders, and accessory suggestions. After a sale, associates keep files on customers, noting style and color preferences, sizes, and other pertinent information. Sales associates follow up with notes or telephone calls if new merchandise seems appropriate for their customers. If a good rapport develops, customers are likely to return.

Sales incentives

In order to improve customer service, many stores are now offering their sales staffs incentives to increase productivity. These incentives, which may take the form of higher salaries or commissions, help to attract better sales personnel. At Younkers in Des Moines, Iowa, for example, salaries are now based on the productivity of each sales associate. At Nordstrom, the salesperson who sells the merchandise is directly responsible for the customer's satisfaction. Nordstrom offers their sales associates a 6.75 percent commission as an incentive.

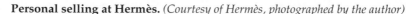

Personal selling at Hermès. *(Courtesy of Hermès, photographed by the author)*

Cooperative salesperson programs

Vendors become directly involved in retail selling by contributing to the salaries of sales associates through various types of programs. Some vendors share the cost of salaries; others pay entire salaries of exclusive salespeople recruited either by the vendor or the retailer.

MERCHANDISING EVALUATION

Evaluation of the success or failure of a season helps buyers plan for better merchandising in the future.

By showing exactly what is being sold, computers help retail executives evaluate a selling season and make decisions about future planning and buying. With the help of this information, they try to measure the merchandising impact on the store's profitability. The analysis can be an ongoing or periodic process.

By comparing current sales of a particular style with the previous week's or month's figures, a buyer can determine whether sales are rising or declining. If they are on the rise, then the style is growing more popular and more of that look should be ordered. If on the decline, those items should be marked down for clearance. Retailers are speeding inventory turns by marking down clothes faster in order to clear out stale merchandise. Yet buyers want goods to "sell through" (at regular price) and prefer to avoid markdowns.

The buyer may compile a *vendor analysis* showing the initial markup, markdowns, and resulting profit percentages gained on products from an individual resource. Although a vendor may have one or two bad seasons, the store will most likely drop a resource whose merchandise consistently gives them no profit.

Sales data can also be very revealing in relation to other important information. When compared with the previous year's data for the same period, the numbers show whether the store's volume grew over the year. When compared with the merchandise plan, the results show how well the store met its objectives established at the beginning of the period, and therefore how well planning was done. When compared with inventory on hand and on order, as well as with customer returns, the figures can indicate whether merchandising, planning, and promotional activities were successful.

Sales data are also used to determine important ratios such as average gross sales, sales per square foot, and stock or inventory turnover.

Average Gross Sales

Total dollar sales for a given period divided by the number of transactions for the same period provide *average gross sales*. This ratio indicates the dollar volume per transaction. An increase in the ratio can indicate higher prices, which result from better merchandise, inflation, or larger sales per customer.

Sales Per Square Foot

Total sales for a period divided by the number of square feet of selling space of a floor or store provide *sales per square foot*, an internationally used indicator of productivity. In their desire for internal growth, many stores are focusing on raising dollars per square foot in existing units, often through renovation. As a result, new departments may be created or existing ones may have their display space increased or decreased.

Stock or Inventory Turn

Stock turn is basically the number of times the stock is sold out and replaced in a given period. A high turnover is usually very desirable because it speeds the buying and selling cycle and constantly frees funds for renewed use, thus increasing the profitability of the retail operation. A high turn reduces the inventory risk; it increases the choice and selection available to the customer, since new merchandise arrives in the store continually; and it reduces the risk of losses due to outdated styles and resulting markdowns.

Stock turn is calculated by dividing total sales by average stock or inventory. For example, if seasonal sales total $1 million and average stock is $500,000, then the stock turn is 2. These *sales-related ratios* are not only used to evaluate a department or store, but are also very valuable in comparing different stores within a retail chain or in analyzing the success of the competition.

All these indicators are used by buyers and management not only to measure the success of the retail merchandising operation but also to refine and improve the long-range planning of future merchandising activities. Evaluation and the establishment of new and more accurate goals will result in a healthier, more profitable retail operation.

SUMMARY

As merchandisers, retailers must buy the right goods and have them in the store when the customer wants them. Merchandising is guided by the fashion director and merchandising managers; responsibilities are divided between merchandising and operations. Store policies and long-range planning help retailers do their jobs effectively. The merchandise plan allocates specific amounts of money for the purchase of fashion assortments. In a fashion assortment, related merchandise is balanced between variation and consistency. Assortment planning is expressed as a buying plan that includes descriptions of the types, quantities, prices, and sizes of merchandise needed.

Buyers' responsibilities cover both the buying and the selling aspects of retailing. Buyers plan what to buy, search the markets for goods that will meet their customers' needs, promote sales, and supervise merchandising and selling, only to begin the cycle again for a new season.

Computers aid in the planning, monitoring, and evaluation of retail sales. Quick Response uses electronic data interchange (EDI) to eliminate delays in the supply pipeline, reduce inventories, and help retailers respond more readily to consumer demand.

Customer service has become very important as have sales training and incentives. Customer acceptance is the basis for success and profitability in retailing. Retailers face a constant challenge to serve the public more efficiently and effectively in order to ensure continued growth and development.

CHAPTER REVIEW

Terms and Concepts

Briefly identify and discuss the following terms and concepts:

1. Fashion director
2. Classifications
3. Merchandise plan
4. Assortment planning
5. Stock turn
6. Open-to-buy
7. Vendor
8. Corporate buying
9. Buyer as editor
10. Line buying
11. EDI
12. Associated buying office
13. Private label
14. Terms of sale
15. Central receiving
16. Preticketing
17. Markup
18. Unit control
19. Stock overages
20. Sales incentives
21. Customer service
22. Markdowns
23. Sell through
24. Sales per square foot
25. Vendor analysis
26. Merchandise representatives

Questions for Review

1. How are merchandising responsibilities organized?
2. Why is planning important?
3. Name five factors considered in planning and buying.
4. Describe the buying/selling cycle.
5. What are a buyer's responsibilities?
6. Why must a buyer research fashion trends?
7. What are the services of a buying office?
8. Describe several aspects of inventory control.
9. Name five customer services.
10. Discuss sales techniques.
11. Why is evaluation important?

Projects for Additional Learning

1. Visit the contemporary sportswear department in a retail store and examine the merchandise. Is it all from one manufacturer, or has the buyer mixed garments from several vendors to carry out a fabric or color theme? Who are the major vendors in the department? What are the price ranges for jackets, sweaters, skirts, pants, and shirts? Which manufacturers have the most innovative looks? Ask salespeople which groups are selling the best and why they think so. Does the department have a good selection of sizes, colors, and styles? Write a critique of the department.
2. Shop a large department store. Study the three different ways to merchandise the suit look. Compare suits in (a) the missy coat and suit department (the suit is priced as a single unit); (b) the sportswear department (components are priced separately); (c) the dress department (components are priced as a unit, usually a dress with a jacket). What suit styles do you find in each department? Compare the selection (assortment), price ranges, quality, and fit in one department with those in the others. Which department has the best value and selection?
3. In a local store, evaluate a sale rack. Is the merchandise marked down, or is it a special purchase (garments purchased at a lower price and offered as a promotional event)? Why was the marked-down merchandise marked down? In your opinion, was it because of styling, poor timing, poor construction, poor fit, unattractive colors, or too high a price? Summarize your findings in a short report.

NOTES

[1] As quoted by Pete Born, "Wexner Shares Trade Secrets At FIT," *Women's Wear Daily*, February 23, 1989.

[2] Nicholas Coleridge, *The Fashion Conspiracy*, (London: Heinemann, 1988), p. 267.

[3] Walter Loeb, Retail Analyst and Consultant, Loeb Associates, Interview, April 27, 1992.

[4] Interview, April 29, 1992.

[5] Speaker, "The Customer is Always Right," The Fashion Group International Retail Forum, December 3, 1991.

[6] Interview, June 23, 1992.

[7] As quoted by David Moin, "Room at the Top," *Women's Wear Daily*, February 16, 1993, p. 18.

A Donna Karan New York vignette at I. Magnin. *(Courtesy of I. Magnin, photograph by Alan Berliner)*

15

RETAIL FASHION PROMOTION

CAREER FOCUS

Fashion promotion provides a wide variety of interesting and creative career opportunities. The advertising director, creative director, ad manager, and copy chiefs supervise artists, production artists, writers, media buyers, merchandise coordinators, and camera operators. Directors of fashion, special events and/or public relations have regional managers or store coordinators to carry out responsibilities in individual stores. The visual merchandising director of a chain of stores oversees regional or individual store visual merchandising managers and a staff of designers.

CHAPTER OBJECTIVES

After reading this chapter you should have attained competence in the following areas:
1. Understanding of the purpose of sales promotion.
2. Explaining the purpose, goals, and procedures of advertising, publicity, special events, fashion shows, and visual merchandising.
3. Describing various types of media and their relation to store needs and target customers.
4. Describing the buyer's role in promotion.

At each level of the fashion industry, the objective is to increase market share and sell products to consumers. *Promotion,* in the broadest sense, is the effort to further those sales by means of advertising, publicity, special events, and visual merchandising. The main purpose of all phases of fashion promotion is to generate more sales by inspiring current customers to buy more and by attracting new customers. The primary challenge of promotion is to get the customer into the store.

Promotion involves communicating a store image or the existence of a product to consumers. It is an attempt to attract the type of customer for whom the merchandise is intended. Therefore, before beginning sales promotion efforts, each fashion business must determine its needs and objectives.

This chapter deals particularly with retail fashion promotion including planning, advertising, public relations, special events, and visual merchandising. The methods used by retailers to promote fashion vary considerably. Merchants must choose the approach best suited to their customers, merchandise, and the size of their business and budget.

PLANNING AND DIRECTION

The sales promotion director, fashion director, advertising and creative directors, visual merchandising director, merchandise managers, and buyers work together to plan promotions.

In small stores, a single person may handle all promotional activities with the help of outside consultants or agencies. In a large store, a sales promotion director manages or coordinates the joint efforts of advertising, special events, visual merchandising, public relations, and the fashion office. The sales promotion director, fashion director, advertising and/or creative directors, visual merchandising director, merchandise managers, and buyers must agree on what to promote, when to promote it, and how to reach their target market. At planning meetings, managers discuss how to communicate fashion trends and important designer promotions within the context of the store's image.

Retailers must communicate a consistent store image and fashion message to customers through various forms of sales promotion, including advertising, special events, and visual merchandising. Stores schedule promotional activities for seasons and holidays throughout the retail calendar year as an integral part of the year's merchandising plan.

In spite of all the scheduling, however, new promotions can supersede all of the best laid plans. At I. Magnin, for example, all plans for Father's Day promotions were cancelled in order to launch Calvin Klein's new men's fragrance. Sales promotion has to be flexible in order to be current.

FASHION ADVERTISING

The advertising director, creative director, art director, writers, and artists work together to create retail advertising. They may be assisted by similar personnel at advertising agencies.

The largest portion of the sales promotion budget (approximately 1 to 3 percent of sales) in a retail store is normally allocated to advertising. *Advertising* involves the planning, writing, designing, and scheduling of *paid* announcements designed to attract customers' attention to a fashion product or event.

I. Magnin art directors Rob Corder and Carol Watts check out the components of a newspaper ad. *(Courtesy of I. Magnin, photographed by the author)*

Advertising style must be altered to reach various types of customers. To the contemporary customer, trendy clothes are sold with sex and sizzle; to the upwardly mobile professional, merchandise is presented in enhanced status images; to the family-oriented consumer, fashion is presented in an atmosphere of hearth and home.

Kinds of Advertising

Stores use three basic types of advertisement: image, item, and promotional advertising.

Image Advertising

Image advertising focuses on fashion image, fashion leadership, community goodwill, a new or remodeled store, or a special event. While it may show merchandise, the goal is to build consumer confidence, community goodwill, create a mood, or create excitement about a new store or event. The A/X Armani Exchange was introduced with dramatic ads picturing nuts and bolts to emphasize its basic merchandise.

Item Advertising

In today's lean economic climate, advertising must sell merchandise. Therefore, many fashion retailers have abandoned image advertising in favor of single item advertising. The goal of item advertising is sales as a direct result of the ad. This type of advertising is favored by vertical retailers such as Ann Taylor, and by manufacturers and retailers involved in cooperative arrangements.

Promotional Advertising

Promotional advertising is price directed. It might proclaim that a store has low prices or it might announce storewide sales or clearances. Some stores are more promotional than others and, therefore, do more promotional advertising. Macy's, for example, advertises many promotional events such as its White Flower Day sales.

Scheduling and Planning

Ads and promotional events are sometimes scheduled up to a year ahead of the selling season, whereas the actual content may be planned closer to the selling season. The *advertising plan* is based on past experience, present conditions, and future expectations. An advertising plan is a guide for a specific period (a week, a quarter of the year, or a season) and for the amount of advertising that a store intends to do in that period to attract customers. A *budget* is prepared, indicating the allocation of funds for advertising production and media. Advertising space in newspapers or time on radio or television must be contracted. A *timetable* is developed detailing how, and by whom, the ad production is to be carried out in order to meet media deadlines and other requirements.

Media

In advertising, *media* is a general term used to cover all methods of transmitting a sales message. The media include newspapers, magazines, radio, television, billboards, displays, and direct mail. To be effective, fashion advertising has to be noticed. Generally, fashion advertising is visually oriented, the philosophy being, "Why tell it if you can show it?"

Advertising-department or agency media buyers must choose the medium that will best promote an event or most likely reach a specific target market. Each consumer group has unique tastes, ideas, and interests, and consequently responds to different media. Media buyers must choose which particular radio or television station, newspaper, or magazine will reach the appropriate customer for specific merchandise.

Media buyers also determine the best possible combination of media to reach a particular target market. Several ads are usually placed in different media to support each other and strengthen the campaign. To balance the spot radio commercial aimed at the car commuter, the store might also place an ad covering the same material in the newspaper, for bus and train commuters to read.

Repetition and *consistency* make the advertising message memorable. The same ad heard at the same time every day on the radio, or a fashion ad on the same page every day in the newspaper, make people keenly aware of the store or brand name as well as the fashion message.

Newspapers

Newspaper advertisements, or *run of paper (ROP)*, are popular among most fashion retailers because

1. They provide visual as well as verbal means of telling consumers what merchandise the company has to offer.
2. They may be offered daily.
3. Layouts, art, and copy are relatively easy to produce.
4. Media costs are comparatively low.
5. They provide a quick turn time from idea to appearance. An idea may be presented at 3:00 P.M. one day and appear in the next morning's newspaper.

Media buyers buy space in newspapers that reach the store's potential customers. Day after day, readers in specific geographical areas or trade groups are exposed to the company's fashion message. The producer or retailer can even buy a certain desirable position in the paper, such as the back page of the front section, often paying premium rates to have ads appear the same place each day. For example, Macy's California advertises daily on the front page of a particular section of the *San Francisco Chronicle*.

The use of preprinted advertising, especially in magazine format, inserted into newspapers is an increasingly successful form of advertising. More stores are using the glossy magazine format because, like a catalog, it has a longer "coffee-table life." Also, vendors, who contribute money toward the cost of advertising, like the huge circulation and relatively low cost per thousand. Circulation can be anywhere from a million copies to many millions for stores spread over many states. Inserts are particularly cost effective when the retailer has many branch stores in an area served

A Saks Fifth Avenue advertisement. *(Courtesy of Saks Fifth Avenue)*

by just one newspaper. The production schedule for a magazine insert is kept very tight (about four weeks for production) to keep pace with consumer demand.

Magazines

National retailers like Saks Fifth Avenue regularly advertise in fashion magazines because they can benefit from national circulation. Magazines also have regional editions that carry pages of local store advertising. Metro or area magazines such as *Philadelphia* and *Southern Living* are used by stores whose target customers live in that metropolitan area or region.

Magazines intended for the general public are referred to as *consumer magazines*. Each magazine aims its editorial fashion coverage at a particular segment of the market, just as stores do; therefore, advertisers use the magazine most likely to reach their target market.

In order to provide direct consumer response, some retailers such as Bloomingdale's insert pages of cataloglike advertisements into magazines. Frequently these ads provide 800 numbers to facilitate ordering.

Television

Television advertising is growing in popularity. The medium offers the advantage of being able to show how clothing might fit into a real life situation.The major drawbacks to TV advertising are the costs of air time and production.

Time buys.

Purchased by media buyers, the cost of *air time* is determined by the length of the commercial, the time of day (prime time, the most expensive, is between 7:00 P.M. and 11:00 P.M., when the greatest number of adult viewers are watching), the day of the week it is to be shown, and the size of the market (it is much more costly to purchase air time in a large city with a large viewing audience than in a smaller city or town). Media selection is based on how successful stations are at reaching viewers as measured by subjective *gross rating points* (1 rating point = 1000 viewers).

Production.

A TV commercial might cost anywhere from $25,000 to $150,000 for a 30-second production. Production costs cover fees for the writer, producer, director, talent, voice-over announcer, camera work, location, editing, music, and expenses for wardrobe, props, lighting, film, and developing. Commercials may be produced by broadcast agencies or by local TV stations.

National television.

National television advertising is used by large national firms that sell and distribute apparel across the country and have large enough advertising budgets to pay for production and buy network television time. J.C. Penney, for example, uses TV advertising in its campaign to become the "nation's largest department store." A network television show reaches millions of potential consumers throughout the country.

Cable television network shopping.

Cable television shopping networks such as QVC and HSN have also become popular vehicles for fashion advertisement and provide an opportunity for direct sales response.

Radio

Because viewers can see the fashion, television has an advantage over radio for apparel advertising. However, radio can make the listener aware of a store location or brand name. Peak radio listening hours, which cost more than other periods, are during drive time, the rush hours when commuters listen most to their car radios. Stations are selected for their target-market appeal. If the target market for a particular campaign is professional men and women, they might be reached most effectively by spot radio ads on a news program at rush hour, when they will be commuting to or from work. A radio spot would reach teens as well, but a rock music program would be a better choice.

Direct response

In order to promote instant sales, there is an increasing trend toward providing the means for direct response in all types of advertising. Fax and 800 numbers are provided in the ad to give the consumer easy access to ordering.

Direct mail

Direct mail is a highly effective form of advertising because it is addressed to each potential customer. Direct mail includes catalogs and statement enclosures.

Statement enclosures are advertisements, often provided by manufacturers, which are sent with monthly billings to charge customers.

Catalogs have become a tremendously popular form of advertising because of the shopping convenience they provide (see the discussion of mail-order retailing in Chapters 12 and 13).

Image brochures are sometimes mailed or given out to customers in the store. These are not catalogs per se, as they do not include order forms, but simply communicate a manufacturer's or store's merchandising image to customers.

Many stores mail specific statement enclosures or catalogs to target audiences such as mothers of small children, large-sized customers, or petite women. Stores can also mail to specific zip codes or to specific active or inactive accounts by using charge account records.

Billboards

Billboard advertising is sometimes used for retail stores. On a smaller scale, this type of advertising has been adapted to bus shelters, especially in large cities. This method has proven effective for companies such as The Gap, Esprit, and Levi Strauss.

Cooperative Advertising

Fiber producers, such as DuPont, Cotton Inc., and the Wool Bureau, and apparel manufacturers often cooperate financially on advertisements with retailers who feature their merchandise. Co-op allocations are based on a percentage of net sales to the retailer and may provide up to 50 percent of media and/or production costs. This additional money enables retailers to make ads larger or to run them more frequently. The resulting increased advertising volume may also help the retailer qualify for a lower media rate. Many retailers would not be able to advertise merchandise without co-op money from producers.

The Advertising Department

Because they do so much newspaper advertising, some large retail stores have in-house advertising departments. The advertising director supervises three areas: art, copy, and production.

Art

The art department is responsible for *layouts,* sketches of how the ad will look. Art-department or free-lance artists execute the final drawings, or photographers take pictures. The use of photography is currently popular because its realistic qualities lend themselves to promotional or direct-sell advertising. Illustrations are especially appropriate for image advertising, where the creation of an illusion may be important.

Copy

Copywriters produce the written description in an ad. The good writer pictures the customer who is reading the ad and writes to him or her. Sales copy, which stresses value, is quite different from high-fashion copy, which emphasizes glamour.

Production

Traditionally, the production department pastes up all parts of the ad into a mechanical composition according to the layout. It is the mechanical technician's responsibility to see that the finished ad is the same as the original idea and layout. This can now be accomplished by computer. The computer scans a photo, produces type, and eliminates the need for paste-up. The production department is also in charge of *traffic,* the flow of artwork and copy between the store, printers, engravers, photographers, newspapers, and radio and television stations. It is now possible to send one ad via satellite instantaneously to newspapers around the world.

A mechanical artist pastes up a newspaper ad at I. Magnin. *(Courtesy of I. Magnin, photographed by the author)*

Advertising Agencies

More and more stores are using outside advertising agencies to handle projects. Agencies often help when there is too heavy a work load for the store staff or for a special project such as a magazine format insert. There are also many stores that use agencies as a cost-effective way of producing timely advertising without having to retain an expensive in-house staff. Payment is usually handled by a monthly retainer based on the amount of time the work takes, plus commission on all media buys and project fees.

In large agencies, groups of people are assigned to particular accounts under the direction of an account executive, who acts as liaison between the agency and the client. Agencies also use free-lance copywriters, illustrators, and production artists to help them with heavy work loads and special projects. The drawback to using agencies is that it takes time for them to become acquainted with the store's point of view. On the other hand, an agency can offer objective ideas about how to project the store's image.

PUBLICITY

■■■■■■■■

Publicity is usually handled by the directors of fashion, special events, and/or public relations and writers on their staffs.

Publicity is the *voluntary* spreading of information about people, special events, or newsworthy topics through various communications media. There are no media costs for publicity, but for that very reason it is difficult to obtain. It is also considered more prestigious than advertising because it is the result of an editor's choice rather than payment of money. Media editors choose the material they will use because they think it may be of interest to the community. However, the media also dictate how, when, and where the message will be used. Publicity helps promote the sale of fashion merchandise by making a style, manufacturer, retailer, trend, or other aspect of fashion better known to the public.

Press Campaigns

Retailers hope to bring their names to the public eye by calling attention to newsworthy developments in their stores. They may create events such as fashion shows or celebrity personal appearances in order to obtain publicity. The campaign may be handled by the public relations office or the fashion office, whichever is more concerned with the topic or event. Alternatively, the publicity package may be prepared by an agency or consultant. The work of a press campaign consists of assembling a press or publicity package and communicating this information to the media.

A *press package* consists of a news release and photographs. A *news release* is a written statement of the important facts about a person, place, or coming event. The news release is often accompanied by glossy photographs which may be provided by merchandise vendors.

Stores provide the media with information about events or topics in the hope that the media will publicize them. Publicity material may be sent to various media. To create an aura of exciting news, the publicity director may telephone the media editors. To achieve maximum benefits from publicity, public-relations or fashion offices send publicity material to the media whose audiences would be most interested in their message.

News may be approached from various angles to ensure each medium a unique story. Ann Stock, former Director of Public Relations at Bloomingdale's, provided an example of a Bloomingdale's A/X Armani Exchange campaign. She told the news story from a fashion perspective (who Armani is, what the shop looks like, why the clothes are different) to fashion editors and approached the store opening from a business angle (A/X shops opening all over the U.S.) for business editors.[1]

Newspaper fashion editors often use publicity releases and photos to write their articles. Fashion magazines give publicity to retail stores in the form of *editorial credits,* the mention of the store name as a source of merchandise that is editorially featured. Radio and television also give some publicity, especially to their advertisers. Paying advertisers are becoming more demanding of the media in their requests for publicity.

SPECIAL EVENTS

Corporate directors of special events and/or public relations, regional managers, store coordinators, and their staffs need superb organizational skills to carry out a wide variety of special events.

Special events are designed to give customers a specific time and reason to come into the store or to create good will. They represent an attempt to replace or renew the personal customer contact that has been lost in many large stores. When well planned and executed, special events can enhance the store's identity, build customer loyalty, and create a sense of community spirit. The retail store as "theater" also caters to the many consumers who enjoy shopping as entertainment. Michael Gould, CEO of Bloomingdale's, commented, "We are in the entertainment business, and if it's pleasant to be in our stores, people will stay there longer."[2]

Most special events are planned and carried out by a special events director and, in the case of a fashion show, in cooperation with the fashion office. In a flagship store, there are usually separate offices that handle each function: fashion, public relations, and special events. In a branch or smaller store, all three functions may be handled out of one office. The special events office must schedule events in conjunction with the advertising department to support store strategies. It schedules guest appearances in the store and sends press releases and invitations for both in-store and community events. These may include fashion seminars, convention tie-ins, community services, the launch of a new designer perfume, and/or fashion shows. One example of a successful event is Macy's Easter Flower Show in San Francisco or their Thanksgiving Parade in New York City.

Fashion Shows

Fashion shows are special events that communicate a fashion story. The selection and organization of the fashions and model bookings are done by the fashion office while invitations and other arrangements may be handled by the special events department. There are four possible ways to organize a show: a formal show, a department show, designer trunk shows, or informal modeling.

1. **Formal fashion shows** take a great deal of advance planning: booking models and fittings, arranging for a runway, scenery, lighting, microphones, music, seating, and assistants. Clothes are generally grouped according to styling, color, or other visual criteria. Models and music are selected to complement the clothes and set a mood. A designer-centered event can cost from $3000 to $25,000. Because of their cost, these shows are usually reserved for charity events.

2. **Department fashion shows** on a much smaller scale, are produced in-store to generate immediate sales. Usually a platform is set up right in the department which carries the clothes.

3. **Designer trunk shows** are done in cooperation with a single vendor and are a popular way to sell expensive collections. Invitations are sent to the best customers according to records kept by sales associates. The designer travels from store to store with the collection which is usually

Izaac Mizrahi backstage before a fashion show at Saks Fifth Avenue.
(Courtesy of Izaac Mizrahi, photographed by the author)

shown on models in the designer collections department. Customers get to see the entire collection unedited by a buyer and may order from the samples in their size. While some designers and retailers do fifty percent of their total business through trunk shows, others find them time consuming, exhausting work, and have given them up.

4. **Informal fashion shows** are the easiest to produce. A few models walk through the store showing the fashions they are wearing to customers who are shopping or having lunch in the store's restaurant. The models can take their time and customers enjoy asking them questions. This is often done in conjunction with a trunk show or special promotion.

VISUAL MERCHANDISING

The corporate visual merchandising director, store planning director, architects, regional creative directors, and individual store visual managers and designers create the store's visual image.

Visual merchandising, which has become increasingly important in retailing, is the means to communicate a store's fashion message to prospective customers. Visual merchandising is a team effort involving management (including the fashion director and the sales promotion director); store planning (and the architects who plan renovations); merchandise managers and buyers; the visual merchandising director, designers, and staff; the sign shop; and the individual department managers and sales associates. The actual display work is done by the visual merchandising department, and the daily arrangement of merchandise is carried out by the department manager and sales associates.

Store Planning and Design

Visual merchandising begins with the store building itself. Management, including the visual merchandising director, store planners, and architects, want the store image to be reflected in the design of the store. Many retailers are renovating existing stores in order to generate new traffic and increase market share. They are trying to create a warm, friendly atmosphere in the process.

The store is divided into departments by fashion category and/or into designer or brand shops. Traditionally, the main selling area is reserved for cosmetics, jewelry, and accessories. The rest of the store is divided into groups of departments and shops that relate to one another by category or price range. Attention is given to space planning and department location. Some shops, usually trend categories, need prime location to attract attention. Destination departments, containing merchandise such as swim suits, lingerie, or coats that customers seek because they are needed, can be in secondary locations. In addition, department relationships, aisle space and traffic patterns are considered to make the departments and merchandise accessible and to help customers find what they want easily. All of this planning must be consistent with the store's image. As many of the same fashions are available at different stores, presentation is an important means of creating the impression of uniqueness on which the survival of a retail store often depends.

Effective visual merchandising on Boulevard Four at Bloomingdale's, New York City. *(Courtesy of Bloomingdale's)*

Macy's Father's Day display. *(Courtesy of Macy's)*

The store's image must appeal to the customer's lifestyle. For example, attracting the mature customer to a store or department with quality merchandise requires the creation of a warm, comfortable atmosphere. This is often achieved with wood paneling, comfortable seating, paintings, and furniture such as those found in an elegant home. In the case of basic merchandise, the retailer may appeal to traditionalists by presenting a clean, orderly environment with neatly folded merchandise such as found at The Gap or Benetton. In the case of the contemporary customer, a store may present a trendy image carried out with glitz, innovative fixtures, and rock music.

Since the ultimate goal of visual merchandising—and of retailing—is to get the customer to buy, visual merchandising must entice the consumer into the store, effectively present the fashions the store has to offer, and show the customer how to wear and accessorize them within the context of fashion trends.

Seasonal Planning

Seasonal merchandising themes are planned many months in advance in conjunction with the seasons, other store promotions, and arrivals of new merchandise. Planning a theme gives the store focus and a consistent look throughout. Working within this framework, the visual merchandising department makes up a *seasonal calendar* indicating the dates on which specific merchandise is to be featured, and the number and location of windows and interior displays that are to show the merchandise. In some cases, the theme may involve a tie-in with a vendor to introduce a new line such as Anne Klein's men's accessories. The challenge for the visual merchandising department is to comply with the vendor's concepts while preserving the store's identity.

Windows

The visual statements made in store windows or in displays at a store entrance are the customers' first encounter with the store and must effectively and correctly convey the store's image and fashion focus. As marketing executive Virginia Meyer commented, "Store windows as well as interior displays attract, compel, and persuade in a subtle or not so subtle fashion.... A good presentation can and should stop you, get your attention, and maybe even make you smile. In a very broad sense, visual presentation not only helps to sell the merchandise itself but the store [as well]. It becomes a part of the store's personality, and is one of the reasons for returning to a store time and again."[3]

"Windows are a uniquely urban phenomenon," explained Steven Kornajcik, Corporate Senior Vice President for R. H. Macy & Co. "A window is a total environment, a complete statement on its own. It can show humor or be theatrical, anything to attract attention."[4] Windows are usually the most dramatic of the store's visual statements. An attention-getting window can entice a shopper into the store.

Special-event windows tie in to events and promotions or convey the spirit of a holiday season. They create excitement and interest to see more in the store.

Fashion-message windows feature the newest fashion trends and suggest ways to coordinate accessories for those looks. These dramatic windows are created to attract attention and persuade customers to buy a new garment and/or accessory.

Direct-sell windows are used mostly by stores that carry popularly priced merchandise. They show a representative assortment of the store's merchandise accompanied by prices to tempt the customer with a possible bargain.

Interiors

The customer is further exposed to fashion and accessory purchase suggestions by interior displays. Displays may be located near the entrances to the store, at entrances to each floor or department, or placed in the various departments on ledges, counters, or platforms. Interior displays are critical for stores in shopping malls, which may have few or no windows. In this case, visual merchandising must capitalize on the wide store entrance that gives the passing shopper a sweeping view of the selling area, giving entry-area displays great impact.

Display areas positioned in front of each department or shop set the tone for the area and attract the customer. These displays highlight the most interesting merchandise from the surrounding area within the context of the store theme. Joe Feczko, Visual Merchandising Director for Neiman Marcus, explained, "The inspiration always begins with the merchandise. The props, the colors, the backgrounds, are all driven by the merchandise and the item. After that, the scale and the scope, the style and the silhouette

An I. Magnin window with a "French Style" theme. (*Courtesy of I. Magnin*)

need to be in consideration. The question that then comes up is how the display is going to fit into...the environment."[5]

Interior displays may take the form of *vignette* or *life-style* displays (mannequins posed in a scene and dressed appropriately), or *single items* shown on a form or stand. Sportswear is usually shown in groups while evening wear may be shown by individual item. By showing total wardrobing concepts including accessories, displays help to both educate and entice the customer.

Display Elements

The elements used to show or enhance the clothes on display include mannequins, fixtures, ladders, poles, platforms, tables or other furniture, paintings and other wall decoration, tablecloths, banners, posters, counter cards, lighting effects, accessories, and props. Music is often added to help set the mood.

Realistic mannequins are popular because the customer can relate to the way the clothes look when worn on the body. Mannequins change with fashion and are made in the image of the current ideal of beauty.

Standards manuals

Stores set standards for consistent visual merchandising. Standards manuals list exact specifications for all display elements, for example, the height of a T-shaped stand and how it should be placed on the floor, placement of tables, and so forth. Quarterly addenda update the standards as new ideas are implemented.

Display packages

The corporate or central visual presentation office may create display packages for branch stores so that the entire group or chain will have the same look. Large display packages are assembled for volume stores and smaller versions of the displays for stores with less space.

In the case of in-store designer or brand boutiques, the vendor has visual merchandising requirements for presenting its line of merchandise. In many cases, the manufacturer supplies the fixtures so that they are the same in every store. It is often difficult to reconcile the individuality of these shops with the store's own identity.

The Department

Visual merchandising must draw the customer to the department most suited to his or her life style. There is no sign that says, "If you are age 38 to 50, live in the suburbs, and wear an average size 10, go to the 'Miss Macy' department." Yet each department is merchandised with certain life-style statistics in mind. Large life-style photographs on the walls, at stores like Wal-Mart, help customers find their way.

Because a customer must be able to see fashion in order to buy it, proper fixtures must be used to display stock. The merchandise itself and the way it is presented must appeal to each specific customer. Stock is arranged in an orderly, attractive manner to contribute to the visual effect of the department and to help customers quickly find what they want. Merchandising and design concepts must be presented correctly: coordinates must be shown together and separates and other apparel displayed in groups by color, fabric, or related items. The department manager and sales staff need to rearrange merchandise in order to give greater visibility to slow-selling goods.

Apparel

In an apparel department, assortments may be displayed on *wall racks*, *rounders* (circular racks), *four-way* or *star* fixtures (with four arms), or *T-shaped stands* (with two arms). Assortment displays must permit customers to see an entire range of colors or styles in each size. To avoid visual monotony, the fixtures are usually mixed. Although wall racks utilize space well and rounders show off a color story, they are not the most desirable means of display because the customer is confronted with nothing but sleeves. On frontal projection fixtures, such as the four-way or the T-stand, the garments face outward so that the customer can see the fronts. The first hanger on a sportswear T-stand is usually an H-shaped hanger showing a coordinated outfit so that the customer can see how the various pieces work together. Fixtures are becoming more sophisticated and many stores are also creating their own custom fixtures to fit in better with overall store design.

The Polo Ralph Lauren shop, at I. Magnin in San Diego, utilizes wall racks, T-stands and folding to display merchandise.
(Courtesy of I. Magnin, photograph by Kim Brun Studios)

Stores such as Benetton and The Gap have made folding a popular and space saving way to display merchandise. Store standards dictate placement of shelves or tables and how the merchandise should be folded and grouped. Retailers have found that merchandise on tables sells better because it is more accessible. Sally Frame Kasaks, CEO of Ann Taylor, remarked, "We were able to increase floor capacity 30 percent by folding and stacking apparel in our casual merchandise areas."[6]

Accessories

Much more attention is being paid to visual merchandising of accessories. Traditionally located in the main selling area, accessories are now also available on the apparel floors so that customers can quickly accessorize their clothing purchases. Accessories can be shown in cases, on shelves or tables, or hung on wall-display racks. Stores must balance the saleability of accessible, "open-sell" merchandise with the security of case display.

Visual-Merchandising Trade Shows

Visual merchandising ideas are presented to retailers at NADI, the semiannual trade show of the National Association of Display Industries and on the West Coast at WAVM, the Western Association of Visual Merchandising show. These ideas must be combined with imagination, perhaps a sense of humor, and developed to give each retail store a unique look.

THE BUYER'S ROLE IN FASHION PROMOTION

Because promotion is a team effort, it also requires the involvement of buyers.

Buyers recommend plans for promotions to the heads of the sales promotion department in conjunction with the projected merchandise plan. Suggestions for advertising, displays, special events, and all other types of promotions are developed with the specialists in these areas, and joint efforts must be coordinated.

Advertising

Buyers request ads on the basis of their merchandise plans. They must provide complete information about the merchandise to the advertising copywriter concerning fabric, colors, styling details, price, and sizes. The garment or accessory itself must be given to the illustrator, layout artist, or photographer. The buyer or assistant buyer must carefully check ad copy for accuracy. He or she must then make sure that the merchandise has been delivered and is on the selling floor with appropriate sign copy when the ad runs.

Visual Merchandising

Buyers may also request window and in-store displays. They choose garments (perhaps with the fashion director) or accessories that carry out the promotional theme and image of the store. They must also make sure that there is a good selection of that merchandise on the selling floor for possible customer purchase.

Special Events

Buyers may initiate special events and fashion shows. For example, the buyer of a designer collection might arrange with the fashion office for a designer to make a personal appearance to introduce a new collection. The advertising and publicity departments announce this event to the community, giving the public the opportunity to meet a well-known designer. If the results are favorable, both the designer and the store may win new customers.

EVALUATION

All people involved in sales promotion: directors, managers, coordinators, artists, writers, designers, and buyers evaluate the effectiveness of their efforts in order to plan for the future.

At the end of a promotional event or an advertising campaign, sales are analyzed and the campaign's effectiveness evaluated. Advertising can be evaluated by sales volume. It is very difficult to analyze sales results in relation to visual merchandising or special events, however. Unless, for example, a fashion show actually takes place in a department and people stay afterward to make purchases, how can the value of the event be measured? Management usually evaluates the effectiveness of a campaign as a whole and makes recommendations for the next year.

SUMMARY

Advertising, publicity, special events, and display help promote retail sales. Advertising is the use of paid time or space in media such as television, radio, newspapers, magazines, and direct mail. There are no media costs for publicity but any material used is the choice of the media editors. Special events such as fashion shows draw people into the store and create community good will. Visual merchandising is the in-store presentation of a merchandising message. Promotional efforts are coordinated with buyers in each merchandising area. Promotion has its limits, however, as it is ultimately the consumer who accepts or rejects fashion.

CHAPTER REVIEW

Terms and Concepts

After reading this chapter you should be able to identify and discuss the following terms and concepts:

1. Sales promotion
2. Advertising
3. Publicity
4. Special events
5. Visual merchandising
6. Life-style displays
7. Image advertising
8. Merchandise or promotional advertising
9. Media
10. Radio and television spots
11. Direct mail
12. Cooperative advertising
13. Layouts
14. Copy
15. Ad production
16. Advertising agencies
17. Press release
18. Editorial credits
19. Inserts
20. Display packages
21. Standards manuals
22. Trunk shows

Questions for Review

1. What is the purpose of fashion promotion?
2. Explain the difference between advertising and publicity.
3. Discuss the types of media used in fashion promotion and give examples of how each one reaches target groups.
4. Explain how cooperative advertising works.
5. Why is it essential for a buyer to be involved in retail promotion?
6. Why is visual merchandising important?

Projects for Additional Learning

1. Find and clip from your fashion-magazine collection five examples of co-op advertising. Find the names of the fiber producer, fabric producer (if mentioned), manufacturer, and stores.
2. Analyze the advertising campaign of a large chain store. Search through newspapers covering a one-month period and clip the store's advertisements. Are they always on the same page of the newspaper? Do they use photography or an artist's illustrations? Do they have a high-fashion image or popular appeal? Do they show the merchandise to good advantage? Do you feel the advertisements are effective? What other types of advertisement do they use? Magazines? Television? Billboards?
3. Attend a special event at a local store. What is the purpose of the event? To draw people into the store? To create community good will? Describe the event and add your own photographs, if possible. Do you think the event was carried out successfully?
4. Visit a local department store and evaluate its visual merchandising. Do the displays carry out a theme throughout the store? Describe the decor and display techniques, both in windows and in interiors. Is lighting used effectively? Is merchandise attractively arranged? Do you feel that the total image of the store successfully relates to the merchandise offered?

NOTES

[1] Ann Stock, Social Secretary for the White House, formerly Public Relations Director at Bloomingdale's, Interview, April 30, 1992.

[2] As quoted by Dianne Pogoda, "Spruce Up to Stir Sales," *Women's Wear Daily*, February 1, 1993, p. 8.

[3] Interview, January, 1993.

[4] As quoted by Dianne Pogada, "At Macy's—Turning Windows Into Worlds," *Women's Wear Daily*, November 30, 1992, p. 12.

[5] As quoted by Rusty Williamson, "Refining Neiman's Vista," *Women's Wear Daily*, Best of Group III/Dallas (May 1990), p. 12.

[6] Interview, April 29, 1992.

Appendix One

CAREER GUIDELINES

How will you fit into the fashion business? Choosing a career—not just a job, but work you will enjoy and build on for the future— is one of the most important decisions of your life. I hope this book will help you make that decision.

This appendix tries to give a realistic picture of fashion career possibilities. It surveys job opportunities in textiles, fashion design, marketing, production, retailing, and promotion. In planning for your career, you should first evaluate your interests, talents and ambitions. Next, apply those abilities and interests to the field that offers you the best employment opportunities.

THE TEXTILE INDUSTRY

If you enjoy working with fabrics, you will find several possibilities for interesting employment in the textile industry. A wide variety of skilled and talented people are needed, including artists to create new designs, scientists to develop fibers and finishes, technicians to develop and work knitting and weaving processes, and salespeople to market fibers to mills and fabrics to manufacturers.

Fiber and Fabric Development

Research

Someone with a background in science or chemical engineering could pursue a career with a fabric company, fiber association, or in the labs of a large chemical corporation which develops new fibers.

Textile Design and Merchandising

Textile designers and stylists need a combination of specialized art and technical training. For print design, designers have to be able to apply their skills to two-dimensional design with the understanding that the end use will be in a three-dimensional garment. Fabric stylists also have to know the technical aspects of fiber and fabric production so they can create interesting new blends of yarns as well as new knit and woven constructions.

Production

The technical skills needed to work in a textile plant must be obtained in an engineering, textile, or vocational college. There are positions in the textile plant for project and process engineers, technicians, supervisors, and managers.

Marketing

Marketing provides a variety of interesting careers including sales, public relations, and advertising. Marketers work with product developers as well as manufacturing and retail customers. Entry-level junior sales representatives can eventually work up to management positions in sales and marketing.

Support Areas

A variety of service positions exist in the areas of finance, human resources, and customer service.

Training and Advancement

Most textile firms are located in the South, although some marketing and styling positions are available in New York City and Los Angeles. Positions for sales representatives are generally found near manufacturing centers.

It is important to try to get some experience—perhaps selling fabrics—before graduation. Some textile producers offer training programs which allow a new employee to experience a number of jobs and get an overview of the company. You might try to create a training program or internship for yourself. Experienced and able people from the technical as well as the marketing end of the business may advance to management.

APPAREL MANUFACTURING

Career opportunities for young people entering the apparel industry vary widely. The most interesting aspect of manufacturing is its diversification. Each person within the company must know something about all areas, so that the company operations run harmoniously. Each of the general branches—merchandising and design, production, and sales—calls for different abilities.

Design and Merchandising

If you have a creative flair, consider becoming a fashion designer. The prospective fashion designer must be artistically creative, yet understand the technical and marketing aspects of the business as well. Besides being responsible for the original ideas for garments, designers must have a thorough knowledge of fabrics, must be able to make patterns, and must understand how a garment is put together. In some companies, designers are involved in every step of the production of the line, from concept to completed garment.

Fashion designing is highly competitive. The better aspiring designers are prepared, the broader their opportunities will be. Graduation from a good design college is essential. Upon graduation, the budding designer might start as an assistant in the design or sample department. Any entry-level job will provide useful experience.

Merchandising offers the opportunity to mesh business acumen with design skills. Retail experience is very helpful as background training. The merchandiser works together with the designer in the planning stages, so that the line of samples will be competitive. The merchandiser usually starts as an assistant merchandiser.

If you are interested in fashion design or merchandising, take every opportunity to observe new trends at fashion shows, visit manufacturers, and learn to do comparative shopping. Read *Women's Wear Daily* and magazines that cover fashion to obtain current information.

Pattern Making

If you are technically oriented, you might enjoy a career as a patternmaker. Patternmakers have an important function in the production process: they translate the design idea into a pattern for the actual garment. A patternmaker must understand basic mathematics, have a good eye for proportion and line, and be a perfectionist.

To prepare for a career as a patternmaker, you must learn how to drape a pattern on a dress form and how to draft perfect flat patterns. Your first job may be as an assistant or as a sample cutter.

Production

There are opportunities for both men and women as supervisors or managers in fashion production. Production managers must be well organized in order to plan and monitor production to ensure that delivery dates are met. A production manager needs good technical and interpersonal skills to get the job done.

An engineering or business education is an excellent background for a career in this area. Only a few large manufacturers have management training programs, so a graduate desiring a career in production should seek any entry-level position available, such as shipping or quality control, just to get a start in the field. Piece-goods purchasing is another interesting area of production.

Marketing

If you enjoy working with people, the field of marketing offers contact with customers and clients. You might begin a career as a showroom assistant or junior sales representative. A college education in merchandising or business and retailing experience are recommended.

Training and Advancement

For the most part, fashion manufacturers are located in large cities. Students interested in design positions must be prepared in most cases to relocate to New York or Los Angeles. However, many companies are now opening plants all over the country, and sales representatives are needed to cover territories from coast to coast.

As most manufacturers do not have training programs, you may try to arrange an internship with a manufacturer on your own. Your ingenuity will be appreciated. Opportunities for advancement in manufacturing are improving as more companies promote from within.

RETAILING

Approximately one of every eight employed persons in the United States works in retailing in some capacity. No matter where you live, a retail store is nearby. This wide availability of employment is the main reason retailing and merchandising majors are offered by so many colleges.

It is absolutely necessary to get retail selling experience while getting a college education. Unless a person has dealt with customers and heard their comments, questions, and complaints, he or she cannot understand the basis of retailing. Working part time as a salesperson at a store near you may provide your initial opportunity and experience.

A few large stores offer summer internship programs for college students. Such an opportunity gives the college student a better chance at employment after graduation. Since most stores do not offer such programs, you may need to ask permission to work as an intern for no pay. Creating this kind of situation for yourself demonstrates your motivation and initiative. Use this time to ask questions and learn as much as you can about all aspects of running the store.

A college diploma is necessary for a retail management career. Retailers tend to hire both business majors and creative majors. They like the creative thinking of the merchandising majors as well as the business skills of the business majors. Many students get M.B.A. degrees to round out their education and give them an edge in the job market.

Training

After college graduation, a fortunate few are able to get into a retail management training program at a large store such as Macy's. A typical program combines working in a variety of departments with formal classes conducted by senior executives and training department personnel. Successful completion of a training program is followed by the opportunity to become a junior executive.

In large stores, there are usually two main tracks to management: one is through buying (merchandising), the other through store operations. Most department-store training programs combine both tracks, providing a well-rounded experience.

Merchandising or Buying Line

The first junior executive position is that of assistant buyer. Assistant buyers aid buyers in all their activities. They spend much of their time maintaining sales and inventory records. In addition, they may act for buyers in their absence.

With experience, the assistant may become an associate or single-vendor buyer, and then a buyer. A buyer might advance to the position of group buyer, divisional merchandise manager, or general merchandise manager. The general merchandise manager is part of top management and sets merchandising policies for the entire store.

Store Line or Operations Track

The second career track in retailing is in store operations. The department manager is responsible for having the goods on the selling floor; keeping current records of stock; and, in a branch store, ordering replenishments of stock from the main store. This person needs a background in sales and management training. The department manager is a liaison among salespeople, customers, and buyers.

The department manager may advance to section manager, floor manager, assistant store manager, operations manager, or manager of a branch store. The operations manager oversees building maintenance, overhead, receiving, and the movement of goods within the store. The store manager is responsible for merchandising, sales, employees, and the general success of the store.

Alternate Training

Most stores now offer flexible career paths from the merchandising track to the operations track and back again. For example, a trainee may begin work as a sales associate, then become an assistant department manager, assistant buyer, department manager, associate buyer, and so on. Able performers from both merchandising and operations might successfully work their way up to top management. Top management is responsible for the administration and organization of the store, establishing store policies, and controlling operations.

Other Specialties in Retailing

Fashion Direction   Some large department stores and specialty stores employ a fashion director or fashion coordinator in each of the branch stores to work on fashion aspects of special events and make fashion presentations to personnel and customers. However, the fashion director's responsibilities are much broader. Fashion directors work with other corporate management to develop the fashion image of the store, travel to market openings, report on new fashion merchandise, and help buyers decide what to order. The position of fashion director generally requires a college education in either merchandising or design and experience in styling, coordination, or buying.

Finance   There are also accounting and financial management positions in retailing for those with an education in finance. Accounting employees keep records of money spent and received, the payroll, taxes, and credit. Financial managers work on budgets and financial control.

Personnel   Other retail opportunities exist in the personnel and training departments, which staff the store with qualified and trained people and handle employee relations.

PROMOTION

If you are creative, you might consider a career in fashion promotion. Fashion promotion informs customers about the goods and services available through advertising, publicity, special events, and visual merchandising. In most cases, a college major in journalism or art is necessary.

Copywriters   Copywriters work on advertising and publicity for all levels of the industry. They may work directly for producers, manufacturers, retailers, or agencies. Fashion writers and editors also work for trade and consumer fashion publications. These positions require top writing ability and some experience, perhaps on a college newspaper.

Artists   Artists do advertising and publicity layouts and illustrations. They usually work for agencies or are hired by them on a free-lance basis. Their artistic skills must be technically perfect. Naturally, commercial art training is necessary.

Visual merchandising designers   Visual merchandising styists decorate store windows and arrange interior displays of merchandise to attract customers. Advancement opportunities include positions as store design managers and corporate design directors.

Special events   Imagination and ingenuity are also needed to direct special events. Experience in staging shows or running school publicity events is a good background.

CAREER RESEARCH

After determining which aspect of the fashion business interests you, investigate prospective employers. There are many different types of companies, both large and small. The advantage of working for a large company is the opportunity for promotion. In a small company it is easier to learn every phase of the business.

Read about companies that interest you. Fashion and business libraries have directories such as *The Fashion Guide, Standard & Poor's Register of Corporations,* or *Dun's Million Dollar Directory* which list addresses and information of national and international companies. Local trade associations also have names and addresses of textile, fashion and retail companies. You can write to these companies for annual reports or other available information.

Attend lectures, fashion shows, fashion group meetings, and career seminars in order to meet industry professionals. Interview them at their offices for class projects. Continually read trade periodicals to keep abreast of industry news so that you will be able to answer questions intelligently at an interview.

THE INTERVIEW PROCESS

Preparations for an Interview

Graduating students often feel defeated before they start the job search. However, it is necessary to be persistent and work at getting a job—prospective employers admire people with drive and enthusiasm.

Résumé

First, you need a *résumé* listing the highlights of your education, experience, and activities. Be selective. List only the courses, experience, and activities that directly relate to your chosen career. Part-time or summer retailing experience is very important in preparation for any fashion career.

Your résumé should be neatly typed on an electric office typewriter or computer. Have it photocopied or printed. You may need to send out as many as 100 résumés to land a job, but don't be discouraged, this is normal. Each résumé must be accompanied by a *letter of introduction* asking for an interview and stating where you can be reached by phone or letter and why you want to work for that company. *Recommendation letters* from teachers and professionals help a great deal.

Be sure to follow up the letter and résumé with phone calls or they may be ignored. Try to get interviews even if there are no positions open with the excuse that you want to ask some questions about the company. An applicant's enthusiasm and persistence can frequently convince an employer that the applicant would be an asset to the company.

Portfolio

If you are seeking a creative position, you will need a portfolio in addition to a résumé. The *portfolio* should look professional, exhibiting only the best examples of your work. Include any projects that won awards or prizes. A design major might include sketches of new ideas (simple technical sketches on graph paper are fine); practice designer work sheets, including all the information on fabrics, trimmings, and labor (see the example in Chapter 9); large photographs of completed garments worn by professional-looking models; and fabric swatches.

Writers should include samples of their work. Especially important is anything that has been published, even in your local or college newspaper.

The Interview

When you finally have obtained an interview, make the most of it. Learn as much as you can about the image and policies of the company beforehand. Each interview is good practice for the next one!

A good appearance is absolutely essential for a job interview in the fashion field. You are making a visual statement of what you know about fashion and about your own self perception.

Your research will be useful during the interview. Brush up on fashion terminology and read current trade periodicals beforehand. Would-be textile and apparel designers should also have up-to-date information on fabric resources. The interviewer may ask questions such as, "Why do you want to become a buyer (designer, etc.)?" and, "Why do you want to work for this company?" Obviously, without proper research you could not specifically answer these questions.

No matter what your training or college major, companies will be looking for the following:

- Good skills learned in college and on the job
- A well-developed résumé
- A professional-looking portfolio (in design or communications)
- A fashionable appearance
- The ability to express oneself clearly
- A pleasant personality and a positive attitude
- Enthusiasm, self-motivation and a high level of energy
- Awareness and an eagerness to learn
- A willingness to take responsibility

THE FIRST JOB

Your first job after school or college should be considered an *apprenticeship,* a period of learning on the job. If your employer has no training program, try to set up your own apprenticeship so that you can move around and learn all aspects of the business. Try to learn as much as possible. Think of your first job as paid schooling. Never stop learning and your career will always be rewarding. Odile Laugier, Vice President of Design at Adrienne Vittadini told me, "I've been here eleven years and I'm still learning."[1]

First, obtain experience. Then demonstrate how efficient and talented you are by doing a good job. Companies are always looking for responsible people to promote to better positions. If you are not promoted, at least you have gained experience for moving on to something else.

Contacts that you make on the job will be valuable later. For example, fabric sales representatives often hear of job openings with apparel manufacturers and spread the word.

Be flexible and pleasant with your co-workers. A little humor and diplomacy go a long way toward promoting positive working relationships. Changes in fashion make the business exciting, but the creative people involved make it even more interesting.

Students often talk of opening their own businesses after graduation, and it seems that working for oneself would be easy. However, many small companies fail due to lack of experience. Before opening your own business, get as much experience as possible, both in a large company and in a successful small one. Working for yourself is very difficult because it requires self-discipline, self-motivation, a large capital outlay, and many risks. There are rewards, of course, for those who have ambition and creativity and are willing to work long hours.

If you want to move up in the fashion field, give your education, training, and work all the effort and enthusiasm you can. Make the most of each situation. You get out of life what you put into it. Best wishes for a successful and rewarding career.

NOTES

[1] Interview, April 30, 1992.

PROJECTS TO PREPARE FOR A JOB INTERVIEW

1. Write your résumé. Include information on education, awards, experience, and interests that directly relate to your chosen field.
2. What fashion career do you think would bring you the greatest satisfaction? List the positive and negative aspects of this career in two columns on a sheet of paper. Analyze why you think you will do well and be happy in this career. What attributes do you have that you could bring to the job? How is your education preparing you? How will you enter the field, and what are your advancement expectations?
3. If you are a design or communications major, outline what you plan to include in your portfolio. Ask your teachers to help you select your best work. Develop an overall graphic theme for your portfolio.
4. Arrange to interview a professional in your chosen field (designer, retailer, sales representative, etc.). Ask about all aspects of the job. What does he or she like and dislike about it? What makes it interesting? How did he or she start out and advance? What valuable advice can this professional give you? College interviews and contacts often lead to jobs.
5. Research a company that interests you. Write to them for an annual report or other information that might be available. Check out the *Guide to Periodical Literature* for articles about the company in publications. Read the material and make notes of important information you want to remember for an interview.

Appendix Two

THE BUSINESS OF FASHION

We use this appendix to briefly discuss business organization, company ownership, and financing because the information applies to all levels of the fashion industry. This is very basic information intended to answer questions you may have while reading the text or reading articles about fashion companies.

FINANCING

To start a textile, manufacturing, or retailing business takes a large amount of capital. *Entrepreneurs* may invest their own money into their businesses but may need more capital. Established companies also need capital to expand. Entrepreneurs turn to investors, banks, finance companies, and factoring firms for financing.

Investors

Entrepreneurs must demonstrate the value of their business or product in order to attract investors. Investors provide capital for a new business or for growth in exchange for partial ownership and perhaps financial control of the business.

Bank Loans

Before a bank lends money to a new company, it is likely to ask for a personal guarantee such as a piece of property as collateral. This type of loan is called a *secured loan* which means they must be backed by valuable equipment or property. *Unsecured loans* are usually reserved for well-capitalized companies with a history of good performance.

Factoring

Apparel manufacturers have a unique method of financing available to them called *factoring*. Manufacturers need money to buy fabric and trims and pay for labor and overhead months before payment from retailers. A factor can provide money to tide them over until manufacturing is complete, goods are shipped to the retailer, and paid for.

A manufacturer can also sell its *accounts receivable* (unpaid orders) to a factor for a cash loan or for credit protection. Factors charge a commission to cover collection services and to protect against credit losses. They also charge interest on cash advances.

TYPES OF OWNERSHIP

When a new business is formed, a decision must be made about its type of organization. Three basic types of business ownership are common in the United States: sole proprietorships, partnerships, and corporations. Each has different criteria regarding taxes, management, liability of the owner, and distribution of profits.

A *sole proprietorship* is a business owned by one person, a *partnership* is a business owned by two or more persons under a contractual agreement; and a *corporation* is a separate legal entity formed in accordance with state laws.

A business often begins as an individual proprietorship or partnership and may change its status to that of a corporation after it has grown. It must be remembered that each company has its own unique concepts, methods of organization, and operations.

Sole Proprietorships

The oldest and most common form of ownership is the sole proprietorship. A sole proprietorship is the least complicated business structure to start. No federal government approval is necessary, but local authorities may impose some license requirements. The structure of a sole proprietorship is very flexible.

Administration

All policy, organization, and profits are under the control of the owner-manager. Day-to-day operations are shared with employees. The owner has the ability to make quick decisions to adapt to changing market needs, including expanding or contracting the business.

Finances

The business profit is taxed as personal income. A sole owner can raise money to begin or expand a business by purchasing on credit, borrowing, and/or investing personal funds in the business. However, the sole owner is personally liable for all business debt to the extent of his or her entire personal holdings. Since the owner is solely responsible for all business debt, moneylenders evaluate the personal wealth of the proprietor as collateral for loans.

Partnerships

A partnership is a voluntary association of two or more persons to carry on, as co-owners, a business. At the formation of the partnership, the partners should sign an agreement stipulating intentions and buy-sell arrangements in the event one partner later wants to withdraw. The agreement should also include the contribution of each partner to the business, the division of profit, and the amount of authority for each member. Usually partners bring their individual areas of expertise into the business. For example, in apparel manufacturing, one partner might be the designer, the second a production specialist, and the third a marketing specialist.

Administration

Each partner has equal administrative responsibility and the various operating functions are divided among them. New business policies and concepts need only the partners' oral agreement to go into effect. Many partnerships have dissolved or become nonfunctional over disagreements on basic business policies. For a partnership to work, the partners must be compatible and confident that they are working toward a mutual goal. One serious disadvantage of this form of business is *mutual agency*, one partner can make a decision that also binds the other partner.

Finances

The partners are taxed separately on their individual returns on their share of income or loss. Each partner is fully responsible and liable for all debt and taxes incurred by the business unless it is a *limited partnership* which limits the liability of certain partners to their personal investment in the business.

It is easier for two or more people to raise money for business than it is for one. The combined resources of the partners can be used for collateral when they are seeking investors or applying for a loan. Because of each partner's full liability for business debt, outsiders are more willing to extend credit and grant better terms.

Corporations

A corporation is a separate legal entity. It is regarded as a legal person, having a continuous existence apart from that of its owners. Unlike a partnership, which is dissolved upon the death or withdrawal of a partner, its existence is not threatened by the death or retirement of a stockholder or a director. Ownership in a corporation is represented by transferable shares of stock whose owners are called *stockholders* or *shareholders*. Stock can be transferred from one party to another without interfering with company operations. The ease of disposing of or acquiring stock makes this form of investment very attractive. Both small and large investors find stock ownership a convenient means of partly owning a business enterprise that can be run by professional management.

The corporation has become the dominant form of business organization on the American economic scene, probably because of its efficiency in pooling the savings of many individuals and gathering together large amounts of capital. There are still many more sole proprietorships and partnerships than corporations, but in terms of output and dollar volume, the corporations hold an impressive lead. Virtually all large businesses are corporations.

Administration

To administer the affairs of the corporation, the stockholders elect a board of directors. The directors in turn select a president and other corporate officers to carry on active management of the business. Certain stockholders may participate in the policies and day-to-day operation of the company. They may even control the corporation if they own the majority of stock. Company founders may retain over 50 percent of the stock in order to retain control of administration and operations.

Finances

The main advantage of a corporation over the other business structures is its limited liability status. The company is responsible for its debt only to the limit of its assets. The shareholder-owner can lose only his or her investment.

A corporation is generally in the most advantageous position of all the types of businesses to raise capital. It can borrow by putting up corporate assets as collateral. The corporation may "go public," or sell stock to the general public as a means to obtain funds. Funds obtained from selling stock are called *equity capital.*

Corporations are taxed on their income. Dividends paid to shareholders are considered personal income and are also subject to tax, causing a double taxation. To avoid double taxation, many corporations pay bonuses or large salaries to their executives rather than pay dividends.

Under the Subchapter S classification of the tax code, corporations that meet certain requirements and limitations are permitted to include corporate earnings as part of the individual shareholders income and avoid federal corporate taxation.

Privately held corporations

There has been a recent trend for former owners or management of corporations to buy all the shares in the corporation to gain or regain control of the business. This is done to avoid takeovers by other companies and/or to avoid poor management/board of directors decisions in the interest of short-term earnings. Levi Strauss & Company (apparel) and Macy's (retailer) are both examples of corporations that were bought back by management or previous owners. These buy-backs are often accomplished by so-called *leveraged buyouts.*

CORPORATE GROWTH

A company may expand by growing horizontally or vertically. Horizontal companies expand on the same level of the industry. Vertical companies combine activities on two or more levels of fashion production and/or retailing. Horizontal or vertical expansion may occur in one of three ways: by internal growth, by merger, or by acquisition.

Internal Growth

Internal growth is perhaps the most desirable because it is real growth in terms of creating new products or services and new jobs.

Horizontal growth refers to adding product lines on the same market level. An apparel company could add new lines to diversify its product offerings; a retail store could open new branches.

Vertical growth refers to growing backwards or forwards in the marketing chain. Many designer and brand name manufacturers have opened retail stores to sell their merchandise. An example of this type of organization is Brooks Brothers, the oldest men's apparel retailer in the United States, founded in 1818 by Henry Brooks. A textile producer could manufacture clothing. The Laura Ashley firm prints and dyes fabrics, manufactures clothing, and sells both clothing and fabrics in its own stores worldwide.

Mergers and Acquisitions

A store or company may become part of a corporation by merger or acquisition (being purchased by a corporation). Companies merge into large corporate organizations for a multitude of reasons:

◆ They may wish to take advantage of a large corporation's purchasing power.

◆ It may be cheaper to acquire another company than to develop a new business from scratch.

◆ Diversification to spread risk.

◆ Tax considerations and personal reasons can trigger a business combination.

Some parent companies want to be involved in management and financial control; others are content to allow the acquired retail firm to run autonomously, as long as the bottom line on the income statement shows the required profits. In both cases, it is possible to cut costs by consolidating duplicate facilities and departments such as marketing, purchasing, and accounting.

Horizontal mergers and acquisitions

In a horizontal merger or acquisition, two companies with the same line of business are combined. Retail stores of relatively equal size often merge to pool their resources. For example, Federated Stores includes Bloomingdale's, Abraham & Straus, Burdine's in Florida, Rich's in Atlanta, The Bon Marché in the Pacific Northwest, and Lazarus in the Ohio area.

In other cases, larger companies purchase smaller ones that are in similar but noncompeting areas. For example, The Limited owns Lane Bryant and Victoria's Secret, both noncompeting retailers. Marks & Spencer of England owns Brooks Brothers. They are both retailers that have manufacturing done for their own private labels but are not located in the same countries. Spiegel owns Eddie Bauer and introduced the catalog business to them.

Vertical mergers and acquisitions

In a vertical merger or acquisition, a company expands either forward toward the ultimate consumer or backward toward the source of raw material. A few textile companies, such as Burlington, have expanded through a combination of mergers and acquisitions (as well as internal growth) to bring all levels of yarn and fabric production together in one corporation. It is difficult to keep track of company ownership because it is continually changing as corporations divest themselves of some companies and purchase new ones.

Leveraged buyouts

A very popular form of mergers and acquisitions in the 1980s was the leveraged buyout (LBO). While a traditional acquisition is paid for mostly with cash or stock, the purchase price in an LBO is often 90 percent borrowed money. In other words, the new owners get to control the business by merely putting up less than 10 percent of the purchase price. In order to pay off the debt, the new owners have to either sell part of the corporation or improve profits by cost cutting. The cost cutting is often so severe that it can be detrimental to the business.

Conglomerates

Many large conglomerates have moved into the textile and apparel field. A *conglomerate* is a diversified company that owns significantly different lines of businesses. The parent company might be in the apparel industry and own businesses with unrelated product lines, or it might be in a nonapparel-related field and own apparel companies. There seems to be a trend in the United States away from conglomerates as many nonfashion companies such as General Mills have sold off their apparel and retail companies. Perhaps they realize that it is best to stay away from areas they know nothing about.

BUSINESS ORGANIZATION OUTSIDE THE UNITED STATES

In many other countries the forms of business ownership resemble those in the United States, with variations based on their unique legal, political, social, and economic development. With the recent democratization of Eastern Europe, we can expect to see more Western forms of business there.

International Company Ownership Terminology

	sole proprietorship	*partnership*	*corporation*
France	Propriétaire entreprise individuelle	Société en Nom Collectif, Société en Commandite Simple	Société Anonyme (SA), Société à Responsabilité Limitée (SàRL)
Germany	Einzelfirma, Eigentümer	Offene Handels-gesellschaft (OHG), Kommanditgesellschaft (KG)	Aktiengesellschaft (AG), Gesellschaft mit beschränkter Haftung (GmbH)
Italy	Imprese Individuale, Proprietario	Società in Nomo Collettivo, Società in Accomandita Semplice	Società per Azioni (SpA), Società a Responsabilità Limitata
Japan	Jieigyo, Kojin, Kigyo	Gomei Kaisha, Yugen Sekinin Kumiai	Kabushiki Kaisha (KK), Yugen Kaisha
Spain	Proprietario	Sociedad Colectiva, Sociedad en Commandita	Sociedad Anonima (SA), Sociedad de Responsabilidad Limitada (SRL)
United Kingdom	Sole Proprietorship	Partnership	Limited Liability Company

FASHION INDUSTRY TERMINOLOGY

Learning the terminology of the fashion industry is an important part of a fashion education. By using correct terminology, you show that you are familiar with the business. Many fashion terms are from the French language, since France has long been the capital of fashion innovation. For further clarification, check the index and refer to the text to see how the term was used.

accessories   Articles worn or carried to complete a fashion look, for example, jewelry, scarves, hats, handbags, or shoes.

acetate   A man-made fiber composed of cellulose chains.

acrylic   A man-made fiber made of long-chain synthetic polymer.

advertising   Any paid message in the media used to increase sales.

advertising director   The person in charge of the personnel and activities of the advertising department.

alta moda   The Italian couture.

apparel   Clothing, not necessarily fashionable.

apparel industry   The manufacturers, jobbers, and contractors engaged in the manufacture of clothing (also called the garment business, the needle trades, the rag trade).

artisans   People who do skilled work with their hands.

atelier (ah-tel-yay')   French word for designer workshop. Ateliers are classified as *flou* (for soft dressmaking) or *tailleur* (for tailoring suits and coats).

balance   Visual weight in design.

balance of trade   Difference in value between a country's exports and imports.

base goods   The solid fabric used as the basis for a group of sportswear.

bodies   Garment silhouettes

book inventory   The dollar value of inventory, as stated in accounting records.

boutique (boo-teek')   French word for a small shop with unusual clothing and atmosphere.

branch store   Store owned and operated by a parent store, generally located in a suburban area under the name of the parent store.

brand name   A trade name that identifies a certain product made by a particular producer.

bridge fashion   The style and price range between designer and better.

buyer   A merchandising executive responsible for planning, buying, and selling merchandise.

buying office   An independent or store-owned office that is located at a market center and buys or recommends buys for one chain or for many stores.

buying plan   A general description of the types and quantities of merchandise a buyer expects to purchase for delivery within a specific period.

chain store organization   A group of stores that sell essentially the same merchandise and are centrally owned, operated, and merchandised.

classic   A fashion that is long lasting.

classification   An assortment of related merchandise grouped together within a department of a store.

collection   A group of garments designed for a specific season.

commissionaire (ko-me-see-ohn-air')   Store representative in a foreign city.

commodity merchandise   Standard basic merchandise.

computer-aided design (CAD)   An integrated computer system which aids in designing and patternmaking, used in both textile and apparel design.

computer-aided manufacturing (CAM)   Computerized patternmaking, grading, marker making, cutting, and sewing machines.

computer-integrated manufacturing (CIM)   Computer connection to integrate computer-aided design and manufacturing systems.

consumer   Someone who buys merchandise.

consumer demand   The effect consumers have on the marketplace.

consumer obsolescence   The rejection of merchandise in favor of something newer, even though the "old" still has utility.

contemporary styling   Sophisticated, updated styling originally designed for the age group that grew out of juniors.

contractor   An independent producer who does the sewing (sometimes the cutting) for manufacturers; an outside shop.

converter   A textile producer that buys greige goods from mills and dyes, prints, and finishes them before selling them to a manufacturer.

cooperative advertising   Advertising costs shared by a textile producer and/or a manufacturer and/or a retailer.

coordinated sportswear   Sportswear designed to mix and match.

corporate selling   Selling management to management without the use of sales representatives.

cotton   A vegetable fiber from the boll of the cotton plant; the world's major textile fiber.

couture (koo-tour')   French word for dressmaking; applied to fashion businesses that make clothes to order.

croquis   Original paintings of textile designs.

custom-made   Apparel made to a customer's special order; cut and fitted to individual measurements; opposite of ready-to-wear.

cutter   The person who cuts material during the manufacturing process.

cutting order   Directions regarding quantity to cut, how to cut, and what fabric to use.

cut-to-order   Cut and produce only against orders.

cut-to-stock   Cut and produce based on projected estimates of sales.

demographic studies   Statistical studies of population characteristics.

department store   General merchandise store including apparel, household goods, and furniture.

designer   A person employed to create ideas for garments or accessories in the fashion industry.

design services   Reports and ideas available by subscription to manufacturers and retailers; predictives.

design resource   Any resource from which a designer obtains ideas, for example, trade newspapers, design reports, fashion magazines, museums, historic-costume books, nature, theater, films, fabrics, etc.

direct-mail advertising   Any printed advertising distributed directly to specific prospects by mail.

discount retailing   Low-margin retailing; retailers able to offer inexpensive merchandise by buying in quantity and keeping operating costs low.

discretionary income   Income left after basic necessities have been paid for.

display   Visual presentation of merchandise or ideas.

disposable income   Income minus taxes; a person's purchasing power.

divisional merchandise manager   A person in the middle management of a retail store; executive responsible for merchandising activities of a related group of departments; supervises buyers and assistants.

dollar merchandise plan   A budget or projection, expressed in dollars, of the sales goals of a merchandise classification, a department, or an entire store for a certain period, including the amount of stock required to achieve those sales.

doors   Fashion industry jargon for the number of retail stores at which a particular product is sold.

draping   A method of making a pattern by draping fabric on a dress form.

electronic data interchange (EDI)   The exchange of business data between two parties by means of computer.

electronic retailing   Shop-by-computer retailing.

elements of design   Design ingredients: color, fabric, line, and shape.

ethnic or folk costume   Traditional national or regional dress; often inspiration for fashion design.

fabrication   Selection of the appropriate fabric for a garment.

factory outlet stores   Stores which sell manufacturer's overruns directly to the consumer.

fad   A short-lived fashion.

fashion   The prevailing style of any given time; implies change in style.

fashion cycle   Fashion change; refers to the introduction, acceptance, and decline of a fashion.

fashion director   The fashion expert who keeps an organization current with changing styles and works with designers or buyers to form the fashion image of the company.

fashion editor   The head fashion reporter at a magazine or newspaper; analyzes the fashion scene and interprets it for readers.

fashion forecast   A prediction of fashion trends.

Fashion Group International, Inc.   An international association of professional women in the fashion business; founded in 1931.

fashion merchandising   The planning required to have the right fashion merchandise available in the proper quantity and place, at the right time and price to meet consumer demand.

fashion press   Reporters of fashion news for magazines and newspapers.

fashion retailing   The business of buying fashion merchandise from a variety of resources and reselling it to ultimate consumers at a convenient location.

fashion trend   New directions in fashion styling.

Fédération Française de la Couture   French couture trade association composed of three main membership classifications (each called a Chambre Syndicale) and associated groups of manufacturers and artisans.

fibers   Natural or synthetic strands from which yarns are made.

filament   A continuous strand of fiber.

findings   Trade term for the functional unseen trimmings needed to complete a garment, for example, zippers and elastic.

finishing   The last treatments given to fabrics; the final handwork or final touches done to a garment.

first pattern   Trial pattern made in the design department for the sample garment.

flagship store   Largest and most representative store in a chain organization.

flax   A natural fiber made from the stem of the flax plant and used to make linen.

flexible manufacturing   A combination of factory methods and machinery used to make production most effecient.

franchising   When a manufacturer sells the rights to retail its merchandise.

full-fashioned knits   Knit garments with pieces shaped on the knitting machine.

furnishings   Men's clothing category including shirts, accessories, and item sportswear.

General Agreement on Tariffs and Trade (GATT)   A contract between goverments to provide a secure international trading environment.

generic fiber name   Family name given to each type of fiber.

globalization   The trend for manufacturers and retailers (and all businesses) to expand throughout the world.

grading   Process of making a sample-size pattern larger or smaller to make up a complete size range.

greige goods (gray goods)   Unbleached, unfinished fabrics bought by converters.

gross margin   The difference in dollars between net sales and the net cost of merchandise during a given period.

haute couture   Those dressmaking houses in Paris that belong to the Chambre Syndicale of the Fédération Française de la Couture and meet the criteria to be on its Couture-Creation list (see Chapter 8).

hot item   A best seller; also known as a *runner* or a *ford*.

Ideacomo   Italian fabric producers' trade fair, held each November and May in Como, Italy, followed by presentations in New York.

ILGWU   International Ladies' Garment Workers' Union.

imports   Goods made in a foreign country.

inside shop   An apparel company with its own manufacturing facilities.

Interstoff   German term meaning "interfabric"; international fabric trade fair, held each November and May in Frankfurt, Germany.

issue plan   Production schedule

items   Garments sold on an individual basis.

Jacquard loom (jah-kard')   A loom invented by Joseph Jacquard in France in 1801 that weaves an elaborate pattern (for example, damask, brocade, or tapestry) by controlling each warp thread separately.

jersey   Basic construction of all weft knits.

jobber   A trader who buys from the producer and sells to the commercial consumer.

junior   Size range of feminine apparel; in odd numbers 3 to 15.

knockoff   A copy of a higher-priced style.

leased department   Within a store, a department run by an outside company.

licensing   Giving a manufacturer permission to use a designer's name or designs in return for a fee or percentage of sales.

line   An apparel manufacturer's collection of styles. Also, visual direction in a design caused by seams, details, or trimming.

line buying   Buying lines from reliable manufacturers.

linen   A vegetable fiber from the woody stalk of the flax plant.

loss leader   An item sold at less than the regular wholesale price for the purpose of attracting retail buyers to other merchandise.

lyocell   New type of solvent-spun cellulosic fiber.

man-made fibers   Fibers made from cellulose in plants or from chemicals derived from petroleum, gas, and coal.

markdown   The difference between the original retail price and a reduced price.

marker   A pattern layout put on top of the fabric for the cutter to follow.

market   A group of potential customers, or the place, area, or time at which buyers and sellers meet to transact business.

market driven   Responding to market or consumer needs.

marketing   The process of planning, promoting, and selling merchandise.

markup   Difference between cost and selling price.

mass production   The production of merchandise in quantity.

media   Means of communication: newspapers, magazines, radio, TV, and direct mail.

merchandise representatives   Consultants trained by manufacturers to train sales associates in the stores.

missy   Size range of feminine apparel in even numbers 6 to 16.

moda pronta   Italian ready-to-wear.

mode   Synonym for fashion; used mainly in Europe.

modular manufacturing   A manufacturing method utilizing a group of people who work together to produce a finished garment.

Multifiber Arrangement (MFA)   A bilateral agreement among exporting and importing nations that provides the framework to prevent import surges.

national brands   Manufacturers' brands that are available nationwide.

natural fibers   Fibers that nature provides: cotton, wool, silk, flax, and ramie.

North American Free Trade Agreement (NAFTA)   Proposed trade agreement to create a free market between the United States, Canada, and Mexico.

nylon   A durable man-made fiber made of long-chain synthetic polymer.

off-price   A price lower than the original wholesale price, or below the normal wholesale price; usually special purchases, closeouts, or overruns.

offshore assembly   Fabric purchased and cut in the United States but sent to Mexico or the Caribbean countries for sewing.

open-to-buy   The amount of money a buyer can spend on merchandise to be delivered within a given period, minus the amount allocated to merchandise on order.

operations   Steps in production; activities of running a business.

overhead   The costs of operating the store or company.

overlock machine   A machine with needle and loopers that creates an edge finish while sewing a seam.

physical inventory   A physical count of stock on hand.

piece goods   The trade term for fabrics.

piecework   Rate by which many factory workers are paid.

polyester   The most widely used man-made fiber, made of long-chain synthetic polymer.

Premier-Vision   French term for "first look". International fabric trade fair held each March and October in Paris.

prestige or institutional advertising   Advertising that promotes a store's image or goodwill rather than specific merchandise.

prêt-à-porter   French for ready-to-wear; literally, "ready to carry."

price line   A specific price point at which an assortment of merchandise is offered for sale.

price range   The range between the lowest and highest price lines carried.

private label   A store's own brand.

production pattern   The final pattern made to company size standards.

progressive bundle system   A manufacturing system requiring one operator to repeat one assembly task and grouping operators to follow the order of production; section work.

promotion   An activity designed to encourage the purchase of a product.

promotional stores   Stores that stress special sales, bargains, and price reductions.

proportion   The relation of one part of a design to another; an important principle of garment design.

publicity   Nonpaid messages about a company and its policies, personnel, activities, or services.

Quick Response (QR)   An attempt to speed ordering and distribution between all levels of the industry via electronic data interchange.

quotas   A means of regulating exports and imports.

ramie   A natural vegetable fiber from the stem of a nettlelike shrub.

rayon   A man-made fiber made from rejuvenated cellulose.

ready-to-wear   Apparel that is mass-produced (opposite of custom-made).

receiving   The area of the store or manufacturer where packages are opened, checked, and marked.

repeat   The repetition of a print in fabric design.

resource   Term used by retailers for a manufacturer, wholesaler, vendor, or distributor. A company that sells goods in the market of finished apparel.

retailing   The business of buying goods at wholesale markets and selling them at retail to the ultimate consumer.

retail price   The wholesale price plus a markup covering the retailer's operating costs and a profit.

sales-per-square-foot   Amount of sales-per-square-foot of store floor space; measure of productivity.

sample   The trial garment or prototype.

sample cut   A three-yard to ten-yard length of fabric used by the design department to make up a trial sample garment.

Savile Row   Street in London famous for men's tailoring.

selected distribution   Limiting the number of stores that may buy merchandise to maintain exclusivity.

sell through   The ability of a line to sell regularly and steadily at full price.

Seventh Avenue   The main street of New York City's garment district; the term is used to represent the whole district.

showroom   A place where sales representatives or management show a line of merchandise to potential buyers; called *salon de presentations* in France.

silhouette   Outline of a garment.

silk   The only natural fiber in filament form; obtained from the cocoons spun by silkworms.

soft goods   Fashion and textile merchandise.

sourcing   Worldwide search for the best available fabrics or garment production at the best price.

spandex   A man-made fiber of long-chain synthetic polymer comprised of stretchable segmented polyurethane; known best by the Du Pont brand name of Lycra.

special events   Activities set up to attract customers to a selling place.

specialty store   A retail establishment that handles narrow categories of goods such as men's apparel, women's apparel, or shoes.

spinning   The process of extruding and hardening man-made fibers; the process of drawing and twisting staple fibers together into yarn or thread.

staple goods   Goods for which there is a demand that continues over many seasons.

stock turnover   The number of times a store's merchandise stock is sold and replaced in a given period.

store image   The character or personality that a store presents to the public.

style   Certain characteristics that distinguish a garment from other garments; a particular look in fashion.

style ranges   Categories of styles that appeal to different consumers.

stylist   A fashion expert; selects colors, prints, or styles for presentation or prepares fashion merchandise for photographic presentation in an advertisement or catalog.

tanning   The process of transforming animal skins into leather.

target market   The group of consumers to whom a producer, manufacturer, or retailer aims products, services, and advertising.

textile fabrics   Cloth made from textile fibers by weaving, knitting, felting, crocheting, laminating, or bonding.

texture   The surface interest of a fabric.

texturing   The process of crimping or otherwise modifying continuous filament yarn to increase cover, abrasion resistance, warmth, resiliency, and moisture absorption or to provide a different surface texture.

toile (twahl)   French word for a muslin sample garment.

trademark   Company's individual registered mark and name for a product.

trend buying   Buying from new resources to obtain fashion newness.

trendsetter   A designer or fashion leader who sets a fashion direction that others follow.

trunk show   Show of designer clothes that moves from store to store, often accompanied by a personal appearance by the designer.

unit control   Systems for recording the number of units of merchandise bought, sold, in stock, or on order.

unit production systems (UPS)   Computer-guided conveyors which move garments automatically from one work station to the next; automatic progressive-bundle system.

universal product codes (UPC)   Standard codes that identify style, color, size, price, fabrication, and vendor on price tags and enable this information to be fed through an electronic data interchange system.

variants   Modifications of basic generic fiber compositions for special applications.

vendor   A seller, resource, manufacturer, or supplier.

vendor analysis   Statistical analysis of the profits made on merchandise from individual vendors.

vertical integration   The joining of companies at different levels of production and marketing, such as a fiber producer with a fabric mill.

visual merchandising   Making merchandise visually attractive to customers.

warp knitting   Knitting fabric in loops running vertically.

weaving   The process of forming fabric by interlacing yarns on looms.

weft knitting   Knitting fabric in loops horizontally or in a circle.

wholesale market   Market where commercial consumers buy from producers.

wholesale price   Price paid by commercial consumers for supplies and products.

Women's Wear Daily   Trade publication of the women's fashion industry.

wool   A natural fiber from animal fleece.

yarn   A continuous thread produced by twisting or spinning fibers together.

INDEX